Organic Materials
in
Aquatic Ecosystems

Author

Humitake Seki, Ph.D.
Associate Professor
Department of Biological Sciences
University of Tsukuba
Japan

CRC Press, Inc.
Boca Raton, Florida

Library of Congress Cataloging in Publication Data

Seki, Humitake, 1937—
    Organic materials in aquatic
ecosystems.

    Bibliography: p.
    Includes index.
    1. Aquatic ecology. 2. Biogeochemical
cycles. 3. Organic compounds. I. Title.
QH541.5.W3S44   574.5′263   81-9928
ISBN 0-8493-6446-9   AACR2

This book represents information obtained from authentic and highly regarded sources. Reprinted material is quoted with permission, and sources are indicated. A wide variety of references are listed. Every reasonable effort has been made to give reliable data and information, but the author and the publisher cannot assume responsibility for the validity of all materials or for the consequences of their use.

Direct all inquiries to CRC Press, Inc., 2000 N.W. 24th Street, Boca Raton, Florida 33431.

© 1982 by CRC Press, Inc.

International Standard Book Number 0-8493-6446-9

Library of Congress Card Number 81-9928
Printed in the United States

INTRODUCTION

Each aquatic ecosystem may not be absolutely independent from neighboring ecosystems in the biosphere, and thus an exchange of organic materials can be expected through the boundaries of ecosystems. The amount of organic materials involved in this exchange is very small when compared to that in the reservoir of many ecosystems, especially in the marine environment.

In such an enclosed aquatic ecosystem, organic materials are produced primarily by photosynthetic activity. Some fraction of these organic materials is transferred, through the grazing food chains, to herbivores and then to carnivores within the ecosystem. The rest is discarded as nonviable organic material into the reservoir of organic debris through biological activity in the food chains. The organic debris is further transformed into viable organic materials through the detritus food chains of the ecosystem. Accordingly, organic materials in the ecosystem are exposed to an endless series of changes which continue as long as organisms continue to live in the system and expend energy by utilizing organic materials.

Professor Timothy R. Parsons inspired the author by presenting stimulating perspectives on the generalized model of the dynamics of organic materials, and by encouraging him to apply ecological principles when the author was a postdoctoral fellow of the National Research Council of Canada from 1967 to 1969. This volume is intended to cover the author's progress, giving more concrete form to these principles. The author also would like to express his gratitude to Professor Claude E. ZoBell, who has always been a source of stimulating advice and encouragement since the author was a graduate student. Professor Parsons kindly has criticized this manuscript and made many useful suggestions for its improvement. Dr. C. M. Lalli has, with great patience, carried through the task of changing the manuscript from Japanese-English to English-English.

Humitake Seki

THE AUTHOR

Humitake Seki, Ph.D., has been Associate Professor in the Department of Biological Sciences at the University of Tsukaba, Ibaraki, Japan, since 1976. Dr. Seki graduated in 1961 from the Department of Fisheries, the University of Tokyo. He obtained an M.Sc. degree in 1963 and a Ph.D. degree in 1966 from the same university. He has been awarded a postdoctoral fellowship from the National Research Council of Canada to work at the laboratory of Dr. Timothy R. Parsons at the Biological Station of the Fisheries Research Board of Canada, Nanaimo, British Columbia.

Dr. Seki was a research associate at the Ocean Research Institute of the University of Tokyo from 1966 to 1976. Since 1966 he has been an educational official of the Ministry of Education of Japan. Also in 1966, Dr. Seki was awarded the Okada Prize from the Oceanographical Society of Japan for his work "Microbiological Studies on the Decomposition of Chitin in the Marine Environment". He is a member of the board of directors of the Oceanographical Society of Japan. He has published approximately 100 research papers and has presented several invited papers at international meetings. Dr. Seki has written two books, *Methods for Microbial Ecology in Aquatic Environments,* Kyoritsu Shuppan Co., 1976 and *Kashima-shin-ryu, the Origin of Martial Arts in Japan,* Kyorin Shoin Co., 1976, both in Japanese. He has also written chapters on aquatic microbiology in two books.

His current major research interests include the dynamics of organic materials and the pollution problems in the hydrosphere.

TABLE OF CONTENTS

Chapter 1*

ORGANISMS IN AQUATIC ECOSYSTEMS

I. EVOLUTION OF LIFE IN THE HYDROSPHERE

The biosphere is defined as the thin film at the surface of the Earth in which life exists. Three conditions are necessary for the existence of a biosphere. First, it is a region where liquid water can exist in substantial quantities. Second, it receives an abundant supply of energy from an external source. Third, within it there are interfaces between the liquid, solid, and gaseous states of matter.

Spores of bacteria and fungi are plentiful in the atmosphere to a height of approximately 1 km, where they are blown by winds from the lower air. They are, however, relatively sparse at heights of 1 to 10 km, with densities of less than $1/m^3$, and most of them are in a dormant state. On the other hand, organisms have been detected on the bottom of the deepest ocean trenches, with depths of more than 10 km. Thus the biosphere has a maximum thickness of approximately 20 km. Active biological processes may take place only within 5 km on land or in the sea, but the majority of organisms live within a zone of only 30 m.

In this very thin layer, the multitude of chemical and biological activities that we call life takes place. Organisms exist there in fine balance by the way in which they interact with each other, depend on each other, or compete with each other, and things could not be otherwise. The balance becomes more obvious when it is disturbed, yet it readjusts remarkably to a new fine balance after a disturbance.

Microorganisms inhabit every place on the surface of the Earth where larger organisms exist. They can inhabit many places where no other life can survive for long. Wherever life exists, there may be microbes; the most extreme conditions tolerated by microbes represent the limits within which life can exist. Microbial activities, conversely, greatly affect such environmental conditions as pH, redox potential, surface tension, electrostatic charges on surfaces, and gas tension.

Through radioactive-isotope dating techniques, it is fairly well-established that the Earth is approximately 4.5 billion years old. The oldest rocks containing fossilized structures resembling bacteria are approximately 3 billion years old. Around 3000 million years ago the Earth had cooled sufficiently for liquid water to exist permanently on its surface. At that time the atmosphere was composed of methane, hydrogen, and ammonium, but contained no oxygen. Ultraviolet radiation is believed to be a potent agent in causing the formation of organic matter from gas mixtures of the kind likely to have existed in the primitive atmosphere. A series of experiments by Miller[1] and others showed that a wet mixture of methane, hydrogen, and ammonium formed traces of organic compounds, including organic acids and amino acids, after exposure to an electrical discharge. Subsequent experiments have shown that all sorts of organic compounds are formed under these conditions, such as the purines which are particularly characteristic of life forms. Before life originated, therefore, the hydrosphere may have been a dilute solution of organic matter formed by electrochemical and photochemical reactions of this kind. Thus the sea and some other parts of the hydrosphere became the type of environment where primitive forms of life could flourish.

The first primitive organism is believed to have been anaerobic and heterotrophic. Its closest living relative may be a nonsporulating anaerobic microorganism without a cell wall. One possible electron acceptor among primordial gases was CO_2, which would have been reduced to CH_4. Another possible anaerobic electron acceptor is sul-

* All figures and tables for Chapter 1 appear after the text.

fate that could have been synthesized chemically by the reaction of H_2S and ozone. The sulfate could have permitted the evolution of organisms such as *Desulfovibrio* which reduces sulfate to sulfide. Other bacteria are known which can biologically oxidize sulfide to sulfur or further to sulfate. Thus a microbial sulfur cycle can go on, even at the present time, without involving higher organisms at all. Such microcosms of sulfur bacteria may have been the dominant biological systems during the early history of Earth. Such a microcosm is called a sulfuretum.

These primitive microorganisms probably had very limited synthetic abilities. They are believed to have consisted of a relatively small number of complex molecules available in the surrounding dilute solution. Since the emergence of life forms would lead to the removal of complex organic molecules from dilute solution, a mutation that enabled an organism to make do with less complex organic materials could give that mutant an advantage over its competitors in favor of development of increased synthetic ability. Thus biological evolution would take place in the direction of simpler nutritional requirements until the first autotrophic organisms appeared.

The earliest kinds of autotrophs might have resembled anaerobes, such as some colored sulfur bacteria that could oxidize sulfides to sulfur and sulfates with the aid of solar energy. There also might have been certain blue-green algae growing anaerobically with the aid of sulfide by reducing CO_2 with solar energy. The emergence of photosynthetic autotrophs generated oxygen from water while converting CO_2 to organic materials. The oxygen concentration in the atmosphere of the primitive Earth rose slowly as a result of oxygen production by algal photosynthesis. Because of the chemical reactivity of oxygen, gases such as ammonium and hydrogen sulfide would have been removed from the atmosphere. Hydrogen would have been escaping continuously into space. Hence, the atmosphere may have tended to consist of oxygen, nitrogen, and carbon dioxide with residual methane, but most of the residual ammonium may have been dissolved in the hydrosphere. Geological evidence shows a drastic change in the reducing atmosphere; iron-containing rocks older than about 2 billion years contain predominantly ferrous iron, whereas younger rocks contain ferric iron. Primitive anaerobic microorganisms should have found this condition very unsatisfactory. The environment then may have favored organisms able to make biological use of the oxygen. An oxygen concentration of 0.2% is believed to have been sufficient for the evolution of respiratory processes. By about 800 million years ago the atmosphere had become fully oxidized by the activities of oxygen-producing photosynthetic organisms.

The emergence of autotrophs, therefore, wrought dramatic changes in the chemical composition of the surface of the Earth millions of years ago. The composition of the atmosphere, soil, and rocks underwent gradual changes to yield the present biosphere. This still is changing slowly, but the average chemical composition of the biosphere has been constant in the last million years. This is largely because every biologically important element can pass through a continuous cycle from the nonliving environment to living organisms and then back to the nonliving environment. Thus almost all chemical changes which occur on the Earth by any biological activity are reversed by some other activity.

After the formation of the biosphere, the sulfuretum was an important system for dynamic biological changes in elements.[2] During the drying up of the Caribbean some 200 million years ago, marine sulfate-reducing bacteria used organic materials for the reduction of calcium sulfate in the sea to calcium sulfide. This was reversely oxidized to calcium carbonate and free sulfur, probably by photosynthetic sulfur bacteria. Thus the sulfur cycle progressed on a large scale to lay down the world's major deposit of sulfur. However, the cycle did not progress further because there was no air available; the drying up of the sea caused organic materials to become concentrated, whereupon microorganisms grew and used up all the dissolved oxygen. The colored sulfur bacteria not only sedimented additional sulfur but also generated photosynthetically other organic materials that may have contributed to oil formation.

At present, a sulfuretum still may be formed on a large scale in places such as the Black Sea, where a thick anaerobic zone exists below 150- to 250-m depth in the central part of the sea.[3] At the transition zone where environmental conditions are changing from aerobic to anaerobic, chemosynthetic activity is predominant. This chemosynthesis is carried out mainly by sulfur bacteria, predominantly *Thiobacillus*, and it is supported by a continual supply of reduced sulfur compounds such as H_2S, S, and S_2O_3 from the anaerobic zone.

Lakes and sulfur springs[4] of this kind exist in various parts of the world. Similar environments also can occur in special marine environments.[5] There are anaerobic microzones in the aerobic water masses of Saanich Inlet, British Columbia, Canada. The nucleus of such an anaerobic microzone may be detritus or aggregates floating in the sea. It has been shown that a greater density of anaerobic microorganisms inhabits the intermediate water mass of Saanich Inlet where there is a lower concentration of dissolved oxygen. Even when seawater from the surface of Saanich Inlet was incubated in a large column, anaerobic microorganisms formed a maximum biomass in one of the microbial stratifications when the seawater sample was more than 60 ℓ. Anaerobic microorganisms could not be detected in surface seawater samples of less than 10-ℓ volume, whereas anaerobes could be found in every 10 mℓ of the intermediate water mass. As the water masses had not become anoxic, these microorganisms must have been in the anaerobic microzone where aerobes and anaerobes were in symbiotic relation.

When a large amount of raw seawater from Saanich Inlet was stored by making aerobic and anaerobic layers in a large column, microbial stratifications were formed; the most active chemosynthesis was produced by the anaerobic *Thiobacteriaceae* in the same manner as in natural meromictic lakes and seas.[4,6] This probably means that the microbial stratification in the anaerobic hypolimnion is formed during stagnant periods by the microflora in the anaerobic microzones, through a process similar to that which was observed experimentally (Figure 1). Such anoxic macrozones usually have been formed by the activity of various microorganisms in neritic regions in aquatic environments as the result of excess primary production and limited circulatory replacement of dissolved oxygen. In such a transient zone of aerobic to anaerobic conditions (i.e., surrounding anaerobic microenvironments), not only chemolithotrophic sulfur bacteria but also mixotrophic microorganisms may grow vigorously.

Such mixotrophic microorganisms as the facultatively autotrophic sulfur bacterium, *Thiobacillus novellus*,[7,8] preferentially inhabit ultraoligotrophic waters of the marine environment with an ecological advantage over obligately heterotrophic microorganisms that depend solely on dilute organic solution. This is because these mixotrophs only require organic compounds for carbon sources while using inorganic compounds as energy sources. The presence of these mixotrophic microorganisms in aquatic environments demonstrates the physiological link between autotrophs and heterotrophs with special reference to the adaptation of organisms to dilute nutrient environments.

Extensive genetic variation and adaptation of eucaryotes could make further evolution possible. However, eucaryotic cells have lost certain procaryotic features such as genetic plasticity, structural simplicity, and the ability to adapt rapidly to new environments. Because of the greater cellular complexity of the eucaryotes, they cannot grow in extreme environments where procaryotes predominate. This defect of eucaryotes is not critical in most parts of the present biosphere because, during the last million years, environmental conditions are in a relatively steady state through the reversal of biological activities.

It is much easier for simple aquatic organisms to live at an interface, especially if one side of the interface is solid. It was suggested by Bernal[9] that the surface properties of solid materials in contact with water were of great importance in the origin and early development of life. Such a beneficial effect also has been shown experimentally by ZoBell and Anderson,[10] who demonstrated that bacteria and other organic materials adsorbed to solid surfaces exhibit a greater efficiency in utilizing dilute organic materials in the marine environment.

Parsons[11] has reviewed the post-Cambrian evolution of marine life (Figure 2). The early pelagic marine environment contained small green, blue-green, and red algae from Precambrian times; dinoflagellates first appeared in abundance at the beginning of the Devonian. The dominant pelagic predators during the beginning of the Lower Cambrian were the coelenterates, including hydrozoa and scyphozoa; the anthozoa appeared somewhat later in the Ordovician. Protozoans and ostracods were plentiful in the Lower Cambrian. Since these animals are largely herbivorous, they could have been among the filter-feeding plankton which filled the step between the primary producers and tertiary consumers in the ancient pelagic sea.

Although the first fish appeared in the Silurian, their feeding habits are believed to have been largely benthic, and the Devonian generally is heralded as the start of the age of fishes. In this period, part of the predatory pattern of the pelagic environment changed from the nonspecific encounter-feeding of the carnivorous coelenterates to the much more metabolically costly feeding of the raptorial fishes. Since raptorial feeders select for larger prey, it is necessary to assume that these became available through a fundamental change in the size spectrum of primary producers. The appearance of the dinoflagellates in the early Devonian brought about an increase in cell size and also an increase in total energy available in the oceans. Representatives of dinoflagellates have cell sizes several orders of magnitude greater than most of the planktonic red, green, or blue-green algae in pelagic regions. They also possess accessory pigments such as chlorophyll c and peridinin which adsorb light more efficiently in the hydrosphere. These changes in primary production permitted the development of a nekton community, which requires more energy because of higher metabolic demands and because nekton require additional energy to pursue their prey by swimming actively.

It has been suggested that evolution of the pelagic crustaceans also occurred during the Devonian, leading eventually to the large herbivorous crustacean community that is so predominant in the oceans today. The abundance of these crustaceans, however, was not simply due to an increase in the size of primary producers resulting from the appearance of larger flagellates. It was the appearance of diatoms in the more recent Cretaceous which eventually allowed the evolution of a third type of food chain in which the top predators were homoiothermic whales. Whales have a metabolic energy requirement some 30 times greater than that of heterothermic fish. With the appearance of diatoms, the primary productivity in certain areas of the hydrosphere was greatly increased because they contain accessory pigments such as chlorophyll c and fucoxanthin, permitting efficient photosynthesis at depth. The diatoms contributed to the later establishment of such important pelagic herbivores as euphausiids in the late Cretaceous and early Cenozoic.

Thus, present marine communities have been able to obtain much more energy with higher efficiencies in energy transformations. With much more energy being made available, higher species numbers and population densities could develop. These have permitted evolution of a greater variety of living organisms, yet microorganisms remain the most important constituents, both qualitatively and quantitatively, of aquatic ecosystems. This huge group of microscopic organisms exists in vast varieties from bacteria to diatoms to protozoans, all similar in structure to primitive procaryotes and eucaryotes.

II. ENERGY FLOW IN THE BIOSPHERE

The biosphere is a dynamic system of chemical changes brought about by biological agents at the expense of solar energy. Solar energy is the primary energy source for photosynthetic organisms in the biosphere and is used indirectly by many nonphotosynthetic organisms through food chains. When a chlorophyll molecule absorbs a quantum of light, the chlorophyll enters into a stage of excitation, i.e., the light energy is present in the excited chlorophyll molecule. The excitation results in an electron being driven off from the chlorophyll. This electron is the energy source, and ATP synthesis occurs during a process called photophosphorylation.

In photosynthetic bacteria, the electron goes around a cycle starting from chlorophyll and ending with chlorophyll, and two molecules of ATP are synthesized. This cyclic process by which light energy is converted into chemical energy of ATP is called cyclic photophosphorylation. No NADP is reduced in this process. In algae and green plants, two cooperative light reactions occur and two ATP molecules and one $NADPH_2$ are synthesized, while water splits with the formation of molecular oxygen. This ATP synthesis with combined light reactions I and II is called noncyclic photophosphorylation, because the electron flow is not cyclic but one-way.

The primary products of the light reactions, ATP and $NADPH_2$, are short-lived and cannot be stored. Photosynthetic organisms circumvent this difficulty by converting carbon dioxide into energy-rich products for long-term energy storage. These products can be oxidized later for the production of ATP. For this purpose, glucose polymers such as starch and glycogen are produced by many procaryotic and eucaryotic organisms. Poly-β-hydroxybutyrate (PHB) is produced by many procaryotes. These polymers often are deposited in large granules within their cells. In the absence of an external energy source, a cell may oxidize these energy-rich storage materials to maintain itself under starvation conditions. Polymer formation has a twofold advantage to the cell. Not only is the stored energy in stable form, but these polymers also have little effect on the internal osmotic pressure of cells. A certain amount of energy may be lost when a polymer is formed from monomers, but this disadvantage is more than offset by the benefits to the cell. In order to be utilized as an energy source, organic compounds must be able to give up electrons and become oxidized. This oxidation must be coupled with a reduction reaction. This oxidation-reduction reaction is coupled with the synthesis of high-energy phosphate bonds in adenosine triphosphate (ATP). The energy of ATP is used in various biochemical reactions.

The potential primary productivity in aquatic environments can be determined as follows. The solar energy intercepted every year by the Earth is 5×10^{24} J. Since the short wavelengths are absorbed by the atmosphere, roughly half of the energy reaches the surface of the Earth. Half of this energy is infrared and cannot be used for photosynthesis. There is a great energy loss by reflection, and 4×10^{23} J/year are available in the ocean as the energy for photosynthesis. Of this energy, 2% is estimated by field observation to be usefully converted into chemical energy through photophosphorylation by marine phytoplankton. On the basis of the Calvin-Benson pathway, 3 mol of ATP and 2 mol of reduced pyridine nucleotides are necessary for fixation of 1 g at. wt. of carbon. The energy used for the synthesis of 3 mol of ATP is 1.2×10^5 J, and the energy required for the reduction of 2 mol of pyridine nucleotides is 3.6×10^5 J. About 0.5 mol of ATP is consumed per mole of carbon in the conversion from carbohydrate to amino acid. One ATP is necessary for every 10 g of dry weight or 2.4 ATP/12 g of carbon; i.e., 6×10^5 J/12 g of carbon are required to convert from carbon dioxide to living materials. Hence, 5×10^4 J are necessary to convert 1 g of carbon from carbon dioxide to living materials. From the energy required to assimilate 1 g of carbon, the maximum primary productivity of the ocean is calculated to be in the order of 1.6×10^{11} tons of carbon per year (Table 1).

On the other hand, it has been determined that phytoplankton can fix annually 100 to 200 g of carbon per m² of ocean surface. The area of the ocean is approximately 3.5×10^{14} m². If this area is combined with a fixation rate of 150 g of carbon per m², the annual productivity is calculated to be 5.3×10^{10} tons of carbon. This amount corresponds roughly to one third of the maximum productivity as calculated previously on the basis of available solar energy.

In the euphotic zone biological processes generally will lead to a net utilization of the nutrient elements (Figure 3). When the rate of utilization exceeds the rate of supply through diffusion and advection, the nutrient concentrations will decrease. These elements are removed from the euphotic zone in the form of phytodetritus and fecal pellets of zooplankton; most plankton cells and fecal pellets are slightly denser than water and sink slowly below the euphotic zone under stable conditions. Because of these mechanisms, the surface waters in most regions are low in nutrients down to the bottom boundaries of the euphotic layer. Hence, most regions of the open ocean are infertile because the inorganic nutrients necessary for phytoplankton growth are present only in low concentrations. On the other hand, many marine coastal regions receive extensive nutrient enrichment from rivers, especially when the river receives high concentrations of organic and inorganic nutrients from sewage and industrial wastes. In other regions where there are divergences, nutrient-rich waters are carried upward (or upwell) from the deep waters toward the surface. Coastal upwelling of water is associated with eastern boundary currents in subtropical regions; this results from offshore surface water transport of currents flowing toward the equator. The upwelling there, is confined to a narrow strip, usually less than 100 km from the coast, and the nutrient-rich water rises to the surface from a depth of 100 to 200 m. The best known examples of upwelling occur off the coasts of southwest Africa, northwest Africa, Peru, northern Chile, and California. There are many other coastal regions where similar meteorological conditions produce upwelling to a much smaller extent. Divergence also may occur in the open ocean, as on the equator and at the northern boundary of the Equatorial Countercurrent. In the countercurrent and the adjacent equatorial currents, four "cells" are present, representing gyrals with horizontal axes, with neighboring gyrals rotating in opposite directions. Within the southern "cell", the water sinks in the regions of the Tropical Convergence and rises at the equator. Within the next "cell", located between the equator and the southern boundary of the countercurrent, the water rises at the equator and sinks at the countercurrent boundary. Within the countercurrent, the water rises at the northern and sinks at the southern boundary. Within the northern "cell", the water sinks at the Tropical Convergence and rises at the northern boundary of the countercurrent.

It has been pointed out that only a small percentage of the sea floor receives sufficient light to support benthic plants. Nevertheless, even in such restricted coastal areas, many large plants have large standing stocks and may be the chief primary producers. In shallower water, such as banks, atolls, continental shelves, and shallow seas, conditions associated with a solid substrate and other regional modifications of the general regime may enable the development of rich populations of large plants. In some areas, benthic plants belonging to a class of algae form the conspicuous offshore growths popularly known as kelp beds. These are the giants among the seaweeds and form the marine forests. These areas receive enough light for photosynthesis, and local water motion enhances the exchange of nutrients and plant wastes. Such areas are estimated to constitute approximately 7% of the total area of the hydrosphere. Thus the quantity of organic materials produced by these attached plants must be relatively large.

The depth to which sunlight penetrates seawater is influenced by latitude, season, transparency of the water and atmosphere, and other factors. There is sufficient penetration of solar energy to support photosynthesis at depths ranging from less than 1 m down to 300 m. Photosynthetic organisms have been confined largely to the topmost 50 to 125 m in the hydrosphere. All species of photosynthetic organisms selectively absorb light of certain wavelengths. Once light has been absorbed by one organism, it is no longer available to another. The absorption spectra of several different organisms measured in vivo shows that some organisms absorb certain wavelengths well and others absorb poorly (Figure 4). These differences are of ecological significance,[14] and the different absorption spectra of various groups of photosynthetic organisms permit aquatic communities to utilize the available solar energy with little loss. The larger plants as well as phytoplankters are distributed vertically on the sea floor or in the water column, progressing from green algae in shallower areas, to brown algae, and finally to red algae down to depths of 30 to 50 m.

On rocky exposed shores, algae dominate the intertidal zone. The most important algae, in terms of biomass and productivity, are the rockweeds, represented in the northern hemisphere by *Fucus* and *Ascophyllum*. In the deeper waters off rocky exposed shores, algal communities are dominated by the genera *Laminaria* (kelp) or *Macrocystis* (giant kelp). These brown algae achieve extremely high biomass densities and production rates. They are distributed mainly in subarctic and temperate waters (Figure 5). Vertical distribution of kelp and giant kelp communities is roughly from low tide level to a maximum depth of 30 m. In a survey of the seaweed zones in Canada,[15] it was found that the zone dominated by *Laminaria* and *Agarum* contributed over 80% of the total biomass. These large brown algae form dense marine forests in which the fresh weight may reach 30 kg/m². In the whole seaweed zone, the average standing biomass was about 4 kg/m². In areas where the intertidal and subtidal shore is sedimented rather than rocky, the sublittoral zone frequently is occupied by monocotyledonous flowering plants commonly known as sea grasses. Such conditions are found in the sheltered bays and inlets of rocky, glaciated shores and especially on unglaciated shores. In temperate waters the most important genus is *Zostera* (eel grass), and in tropical waters there is a multiplicity of genera such as *Thalassia* and *Cymodocia* (Figure 5). In some areas, particularly those with colder and more humid climates, the sea grasses penetrate into the lower part of the intertidal zone. In sedimented areas, particularly in the extensive intertidal deposits of estuaries and behind barrier beaches, salt marshes tend to form and usually make a major contribution to coastal productivity. Very large areas are dominated by luxurious growths of cord grass (*Spartina*) in the intertidal zone and by *Juncus* and *Salicornia* at or above high tide.

Mangrove swamps dominate the world's coastlines between 25° N and 25° S, extending 10 to 15° further south in eastern South Africa, Australia, and New Zealand, and 7° further north in Japan. They comprise a distinct coastal zone dominated by a few species of moderately large evergreen trees.

Solar energy also is utilized by larger plants drifting in the pelagic waters as well as by those attached on the shallow sea floor. These plants are usually accumulations of large shore plants drifting not only in coastal regions but also in special convergent oceanic areas such as the Sargasso Sea. These drifting plants contribute appreciably to the primary production in the marine environment.

In the immediate vicinity of the coast, seaweeds, sea grasses and mangroves are much more productive than phytoplankton. Their rate of carbon fixation on a unit area basis is estimated to be an order of magnitude greater than that of the phytoplankton (Figure 6), with production values ranging from 50 to 2000 g C per m²/year and averaging from 500 to 1000 g C per m²/year for typical coastal macrophytes. These may be compared with the world production averages for phytoplankton; i.e., 50 g C per m²/year for the open ocean and 100 g C per m²/year for the coastal region. The macrophyte fringe of the oceans has an intensity of production which is as much as 40 times that of phytoplankton production in the open ocean. Very little of this production enters the marine grazing food chains;[15,16] almost all enters the detritus food chains which are closely associated with heterotrophic microorganisms.

Therefore it is apparent that these macrophytes, although restricted to coastal regions, make a great contribution to total marine primary productivity. Large brown algae comprising marine forests have the same productivity as reeds in temperate swamps, which annually produce 1200 to 1800 g C per m². The productivity of large marine algae is even almost equivalent to that of the most productive communities in the tropics, where rain forests and perennials under intensive cultivation may produce annually 2000 to 3200 g C per m². Incidentally, this productivity is believed to be the highest in the biosphere.

The area inhabited by larger plants may be small compared with the whole ocean, but it is an area which constitutes a nursery ground for a very large proportion of the commercial fishes of the world because of the sheltered conditions and rich food supply.

III. GRAZING FOOD CHAINS

The energy that sustains all living systems is solar energy. The energy fixed in photosynthesis is transferred through communities by plants being eaten by animals which in turn are eaten by other animals. This transfer of food energy through a series of organisms is referred to as a food chain.

There are two kinds of food chains: a grazing food chain and a detritus food chain. The grazing food chain is defined as starting with photosynthetic plants which are eaten by herbivores and these by carnivores. On the other hand, the detritus food chain begins with dead organic matter which is utilized by microorganisms and then transferred to detritivores and their predators.[17] Both kinds of food chains interact to form the food web of an ecosystem.

In nature, the food and feeding relationships of plants and animals do not comprise isolated linear systems, but instead a large number of food chains interconnect. This interconnection leads to complex food relationships being referred to as a food web rather than a food chain.

In the euphotic zone of an aquatic ecosystem, four different seasonal cycles generally are recognized based on changes in standing stock of phytoplankton and zooplankton (Figure 7). The first type of seasonal cycle is characteristic of arctic or antarctic waters, where the amount of light is only sufficient for a single plankton bloom during the summer. The second seasonal cycle is characteristic of North Atlantic temperate waters, where breeding of zooplankton cannot start until a spring increase in primary productivity has occurred. The third seasonal cycle is characteristic of the North Pacific Ocean where neither the beginning of the zooplankton breeding nor the size of the zooplankton standing stock is dependent on the presence or absence of phytoplankton in early spring. The fourth seasonal cycle occurs in tropical oceans where there is very little evidence for predominant maxima and minima associated with seasonal events. In many of these cases the grazing food chains must be predominant in the euphotic zone, because the increases and decreases of phytoplankton are followed by those of the zooplankton feeding thereon. Such grazing food chains are especially prominent during or just after the seasons of phytoplankton production; i.e., rapidly growing animals incorporate large quantities of phytoplankton substances into their own structure.

The complexity of food relationships in an aquatic ecosystem can be simplified by considering the food web in the Antarctic Ocean, where major food chains are relatively simple. Because of the immense fertility of the water maintained by the upwelling of mineral nutrients, the pelagic region of the Antarctic Ocean is richer in life than any other comparable oceanic area. This is in contrast to the coastline of Antarctica and submerged ledges of the Antarctic islands which are barren of benthos and fish. The offshore benthic communities of mollusks, brachiopods, pycnogonids, echino-derms, corals, tunicates, hydroids, holothurians, and sponges feed on organic detritus derived from the surface of the water column.

The major species in each family of plants and animals native to the Antarctic are small in number and have evolved more recently than those in the tropics. However, the fewer Antarctic species produce higher population densities than species in milder oceans.

The primary foodstuff ultimately is restored to the water in the form of excretion and dead bodies, which are decomposed into inorganic nutrients by heterotrophic ac-tivities. Upwelling then returns these nutrients to the surface layer where phytoplank-ton photosynthesis takes place (Figure 8). Thus the whole process is recycled. The surface water is so rich in nitrogen, silica, and phosphorus that these are never com-pletely utilized by phytoplankton. With a constant supply of nutrients for phytoplank-ters, the ecological pyramid of Antarctic oceanic life remains stable.

Diatoms are predominant among the Antarctic phytoplankters. Most diatom cells are slightly denser than seawater and would sink slowly to the bottom under absolutely quiet conditions. As they sink, they divide and populations in the upper waters are replenished continually from below by turbulent upwelling water (Figure 9). This mechanism operates very easily in the Antarctic Ocean because there is an adequate chance of a lift back to the surface for the cell and some of its descendants (Figure 8). The boundary between the eastward and westward current systems marks the position of the Antarctic Divergence, which is characterized by the upwelling of subsurface water. As the Antarctic Upper Water moves north, it meets the warmer and lighter Subarctic Surface Water at the Antarctic Convergence, where the denser water sinks and mixes with the waters above and below. Therefore, at both the Antarctic Diver-gence and Convergence, nutrients and diatoms are continuously brought up to the surface layer and high productivity is maintained. Phytoplankton are the major pri-mary producers in the Antarctic Ocean, as larger plants are scarce in the intertidal areas of Antarctica. Under optimal conditions of light, temperature, and nutrient sup-ply these phytoplankters multiply to hundreds of thousands of cells per liter of water, forming "phytoplankton blooms"; these may be yellow, green, red, or brown, de-pending upon which species is dominant.

Euphausia superba, the red crustacean called krill, is efficiently equipped for grazing on these phytoplankters; its hind limbs sweep currents of water toward a net of hairlike bristles under the forepart of its body, where diatoms are entrapped and then swept into the mouth. The total annual production of krill has been estimated to be at least 1350 million tons;[20] this averages approximately 100 g/m². The krill are confined to a thin zone within about 10 m of the surface. They swarm in shoals and windrows meas-uring a few thousand square meters in size, and aggregations of such swarms some-times extend for hundreds of square kilometers. Whales browse in their midst, singly or in herds, and consume vast numbers of them. In the Antarctic Ocean, krill form the most important link between the primary producers and the tertiary consumers; krill are the principal or exclusive food not only of whales but also of some species of seals, penguins, many other oceanic birds, and many fishes (Figure 10).

At each step in this food chain, there is a considerable loss of organic matter and energy. In herbivores, as much as 90% of the total food intake may not be assimilated and will pass out of the body as feces. In carnivores, as much as 75% of the food eaten may be assimilated, although 30 to 50% is more normal.[21] Only a small fraction of what an animal consumes is transformed into body tissues. The rest is dissipated as energy in various forms, a large part of it being expended as mechanical energy in the predator's pursuit of its prey. In general, 80 to 90% of the organic matter is lost at each step in the food chain of the Antarctic Ocean (Figure 11).

Energy-flow studies generally are considered to have been initiated by Lindeman,[22] who first drew attention to the dynamics of ecological food chains by defining the various trophic levels and by utilizing the physical concepts of energy content and transfer. Hence ecosystem energetics can be studied by measuring the efficiency with which autotrophs convert solar energy into chemical energy of plant protoplasm and the efficiencies with which this is utilized by all the heterotrophs. The surplus energy appears as environmental heat, generated in respiratory metabolism; this is the thermodynamic consequence of the extraction of useful work from metabolic energy sources. Thus the loss is the payment for the creation and maintenance of ordered structures at both the organism and ecosystem level.

In aquatic environments, it may be assumed for general purposes that the ecological transfer efficiency at the herbivore level is probably no less than 20% and that transfer efficiencies at higher trophic levels probably range between 10 and15%. Ryther[23] assigned a 10% overall efficiency to the oceanic food chain, a 15% efficiency to the continental shelf food chain, and a 20% efficiency to the food chain in upwelling regions.

IV. DETRITUS FOOD CHAINS

All species of phytoplankton are not equally suitable as food for each marine zooplankter, the most obvious limiting factors being the size and shape of each plant cell. There is no doubt that in the euphotic zone most of the phytoplankton generally enter the food chain directly as food for herbivorous zooplankters, mainly crustaceans. This is partly supported by evidence showing a good correlation between the time of phytoplankton blooms and the fecundity and egg-laying timing of copepods. Other supporting evidence is the temporary increase in the standing stock of phytoplankton, followed by a decrease as the grazing pressure from an increased standing stock of zooplankton becomes effective. However, a succession of small increases and decreases in phytoplankton and zooplankton standing stocks may occur independently throughout the year or temporarily in a certain region (Figure 7).

It also has been determined that marine macrophytes generate detritus rather than entering the grazing food chain, and microorganisms are the most important consumers of this detritus. The semidigested plant detritus is often a significant fraction of the particulate organic matter in coastal regions and forms, with its accompanying microorganisms is a valuable food source for animals. Thus detritivores can derive their nourishment mainly by stripping microorganisms from dead or live plant material.[15,24] Accordingly, especially in the disphotic and aphotic zones, dead material and fecal pellets may be reutilized by microorganisms and the process repeated until all the plant material has been utilized completely. Parsons and Strickland[25] speculated on the importance of the detritus food chain in the ocean from a quantitative point of view. The mean standing crop of heterotrophic microorganisms was estimated as approximately 0.1 mg C per m^3 in the oceans of the world between 50° N and 50° S, although the amounts are less in arctic regions. Thus organic debris can be utilized for the sustenance of higher marine animals through the detritus food chains initiated by heterotrophic microorganisms in the pelagic environment. Assuming a value of approximately 0.1 mg C per m^3 for the standing crop of heterotrophic microorganisms and a growth rate roughly comparable to that of the phytoplankters, the heterotrophic biomass production per m^3 is as much as 0.5 to 1% of the photosynthetic production in the euphotic zone. When it is taken into account that the depth of the total water column in the ocean is 50 times or more that of the euphotic zone, the heterotrophic production beneath a unit area of the coastal ocean could well be of the same order as photosynthetic production.

The contribution of fecal pellets from zooplankton has been shown to be important for the transport of organic debris into deeper oceans; i.e., the debris enters the detritus food chains of deep-sea communities. These intact fecal pellets sink fairly rapidly, although once the mucous cover is broken, the pellets break into small particles which sink very slowly but which rapidly lose their organic components by dissolution. Vinogradov[26] showed that only 10 to 20% of dead pteropods still retain some tissue in their shells when they sink below 500 m and no tissue remains in shells collected below 2000 m. Because of the weight of the shell, dead pteropods sink rapidly from the surface waters which they normally inhabit. Since dead copepods and other crustaceans of similar size sink much more slowly than pteropods, they are believed to disintegrate completely at shallower depths.

In fish ponds of the Volga Delta, Kusnezow et al.[27] observed that the rushes, *Scirpus* and *Phragmites*, began to decompose 5 to 6 days after being cut, resulting in the death of carp due to a lack of dissolved oxygen. After this experience, rushes were cut so as to leave a zone approximately 10 m wide in the Jamat fish pond. Bacteria and bacteriovorous zooplankton multiplied rapidly and exhausted the dissolved oxygen in the area where rushes had been cut. In contrast, in the uncut oxygenated rush zone, carp fry gathered and fed on the zooplankton and grew well. This is an example of a detritus food chain involving organic matter, bacteria, zooplankton, and fish in a successful aquaculture system. The same detritus food chain has been utilized successfully for many hundreds of years in goldfish cultivation in Japan. In this latter case, the detritus food chain starts from the organic matter of stable manure, with the upper trophic levels being comprised of *Daphnia* and goldfish.

In the cases discussed above, detritus food chains are the predominant routes by which the chemical energy of detrital organic matter becomes available to the biota.

The major part of organic matter is present in solution in most aquatic environments. As concentrations in natural waters are very low, dissolved organic matter may be kept at this low level by the efficient transport system of nutrient assimilation by heterotrophic bacteria. There is very little evidence that forms other than bacteria make any appreciable use of this low concentration of dissolved organic matter. Especially in the marine environment, the dissolved organic matter averages one order of magnitude greater than the particulate organic matter. The quantity of particulate organic matter occurring as detritus has been estimated to be several times greater than that present in living organisms.

Chemical analyses indicate that this particulate detritus in the marine environment is composed of 30% carbohydrates, 70% of which is crude fiber, and less than 1% lipids.[28] More than 50% of detritus is composed of proteinaceous substances, having a nutritionally desirable amino acid composition for most marine animals.[29,30] These results indicate that from one half to three fourths of the particulate organic matter should be assimilated easily by marine animals as well as by heterotrophic microorganisms, providing that it is capable of being digested.

Among the heterotrophic bacteria isolated from the seawater of Departure Bay, Canada, *Pseudomonas* sp. was used as a food source for the brine shrimp, *Artemia salina*. The bacteria were grown on two different concentrations of Bacto-peptone and on one concentration of water-extracted sediment materials from Departure Bay, equivalent in organic content to 0.005% Bacto-peptone. The results show that the sediment extract could produce a bacterial population that allowed for the growth of brine shrimp up to approximately the fourth instar, while 0.005% and 0.01% Bacto-peptone allowed for growth up to the fifth and sixth instars (Figure 12). No growth was obtained if the particulate sediment was fed directly to brine shrimp, and the survival of brine shrimp in the absence of any food source was less than 48 hr. It is apparent from this experiment that even sedimented organic materials, which may be less nutritious due to a lower organic fraction than particulate materials suspended in seawater, can be returned readily to the food chain through bacterial action.

Detritus is composed of organic debris and viable microorganisms. The organic debris generally forms a matrix in which microorganisms are embedded, and the whole aggregate forms a semi-independent ecosystem from the ambient seawater. Many marine bacteria and allied microorganisms are associated with such aggregates.

Bacteria may themselves contribute to the formation of larger aggregates. The simplest case of aggregate formation is when the bacteria themselves form microbial clumps without a matrix (Figure 13). Although some marine bacteria may form clumps very easily, it was observed that there was a species-specific tendency towards clumping. Microbial hold-fasts are chiefly responsible for the microbial ability to adsorb on a matrix. A typical aggregate from the hydrosphere, taken after a phytoplankton bloom, shows bacteriovorous protozoans as well as a large number of bacteria and phytoplankters inhabiting the inside of the matrix (Figure 14). Large aggregates help to maintain the density of prey for herbivorous filter feeders at concentrations higher than the prey density at which grazing may normally occur. Thus, if threshold concentrations of particulate food are important in grazing kinetics,[33] the presence of aggregates can increase the feeding efficiency of filter feeding animals.

Baier[34] hypothesized that organic debris is resistant to digestion by animals and that, generally, only the microbial fraction of the detritus can be digested. According to this hypothesis, the lowest trophic level of the community within the aggregate microenvironment is formed by heterotrophic bacteria and allied microorganisms; this applies especially in the aphotic zone of the hydrosphere. These organisms assimilate the organic debris of aggregates as well as dissolved organic matter in the ambient water, following Wright-Hobbie uptake kinetics as originally described by Parsons and Strickland.[25] Thus the heterotrophs form an endpoint of a micro-food chain in the aquatic food web.

Baier[34] divided bacteria feeders into three categories:

1. Those which feed directly on bacteria (e.g., rotifers, copepods, ciliates, flagellates, larval stages of additional invertebrate genera, etc.)
2. Those which swallow the entire substrate, digesting the usable portion and rejecting the rest (e.g., nematodes, *Tubificidae, Chironomidae*, mussels, crustaceans, rotifers, ciliates, flagellates, etc.)
3. Those which graze on solid surfaces (e.g., snails, ostracods, copepods, amoebae, etc.)

Bacteria feeders within the aggregate environment are very small and belong chiefly to the first category (Figure 15).

In order to study the feeding relationships and the importance of aggregates, a microenvironment was designed in a continuous culture system (Figure 16) which was based on the assumption that an enriched culture is essentially a device used to enlarge a microenvironment to macroscopic dimensions.[14] In this system, the marine bacterium was a saprophyte with a great tendency to clump in the seawater medium, *Oxyrrhis* and *Uronchia* were bacteriovorous protozoans, and brine shrimp fed on bacteria as well as protozoans. The first flask was used for the pure culture of the bacteria, and the population density was adjusted by the flow rate regulator between the flask and the reservoir of culture medium. The bacterial culture and excess nutrients overflowed into the second culture flask, where the protozoans grew on the bacterial food and formed zoogloeal aggregates which included bacteria. These zoogloeal aggregates, or colonies of bacteria-forming jellylike masses with attached bacteriovorous protozoans, overflowed into the culture vessel containing omnivorous brine shrimp. The highest crop of brine shrimp could be attained when the system was regulated in the following way:

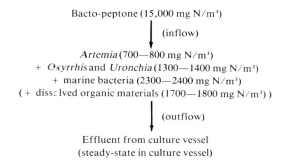

Bacto-peptone (15,000 mg N/m³)

(inflow)

Artemia (700—800 mg N/m³)
+ Oxyrrhis and Uronchia (1300—1400 mg N/m³)
+ marine bacteria (2300—2400 mg N/m³)
(+ dissolved organic materials (1700—1800 mg N/m³))

(outflow)

Effluent from culture vessel
(steady-state in culture vessel)

The standing stock of bacteria — the primary producers of biomass in this detritus food chain — essentially attained a steady-state equilibrium with continual biomass supply from organic debris and continual removal by grazing of detritivores. A model of this detritus food chain resembles an upright pyramid when constructed either in terms of numbers, biomass, or productivity; hence the trophodynamic structure of the system is stable. There are many occurrences of such sporadic microenvironments in the hydrosphere (Figure 15).

The growth of brine shrimp on marine sedimented material has been demonstrated, following the conversion of sediment debris to bacterial biomass.[31] Thus, recently sedimented organic material in coastal environments can readily enter the food chain through bacterial action; however, the quantity of sedimented organic matter utilized by this route is only a fraction of that actually available by whole food webs.

The food chain of bacteria, protozoans and crustaceans can be shortcut temporarily, with the brine shrimp feeding directly on bacteria. The brine shrimp in this case can grow to a certain stage but do not become adult. The intervention of protozoans in the detritus food chain results in an energy loss during ecological transfer to a higher trophic level, but there is valuable compensation in conservation of the top predator, which is Artemia in this system.

Many microorganisms that live in the intestinal tracts of aquatic animals synthesize certain vitamins, which frees their hosts of the need to obtain these vitamins in their diets. On the whole, the beneficial effects of microorganisms in association with higher animals far outweigh their harmful effects.

No aquatic animals are able to obtain a considerable portion of their food from absorbing dissolved organic matter, because the concentration of dissolved organic material is generally maintained at less than 1 mg/ℓ by aquatic bacteria. In such environments with dilute nutrient supply, the detritus food chains contribute a stable food supply to organisms of higher trophic levels since all processes in these food chains should not fluctuate greatly. However, detritus food chains sometimes can be of primary importance in the food web of the euphotic zone, and the processes therein may be somewhat more variable. In St. Margaret's Bay in Canada[36] for example, the spring phytoplankton bloom has been recognized by peaks of large particles appearing in April and May, but nonliving particulate matter composed of detritus produced mostly from macrophytes reaches a maximum in late February and early March. There is little doubt that detritus is of secondary importance much of the year within the euphotic zone, but it may be important as a stable food resource for herbivorous zooplankton during times of low phytoplankton as well as being a food source for many animals which permanently inhabit the deeper regions. On the other hand, the most dynamic processes in detritus food chains can be expected to occur in hypereutrophic environments. Lake Kasumigaura in Japan is one such environment where the population density of blue-green algae during the summer bloom is the highest (682 μg ATP/l) that can be expected in any aquatic system and reflects extreme eutrophication. Total organic matter in this environment was shown to approximate the following distribution in relative units:[37] organic solutes, 100; detritus particles, 200; phytoplankton, 400; and bacteria and allied microorganisms, 60 during the period without an algal bloom. These distributions can be compared with those in oligotrophic oceanic waters:[38,39] organic solutes, 100; detritus particles, 10; phytoplankton, 2; and bacteria and allied microorganisms, 0.2. Due to such enhancement of heterotrophic processes under various degrees of eutrophication, the fraction of organic matter increases firstly in heterotrophic microorganisms, secondly in phytoplankton, and finally in detritus. This precedence of heterotrophic over autotrophic processes is possible because of the small size and large surface:volume ratio of bacteria compared with higher organisms. These factors permit a rapid exchange of substrates and waste products between the cells and their environment. The precedence of heterotrophic processes must be favorable not only for enhancing the detritus food chains but also for maintaining an ecosystem in a steady-state type such as oligotrophic or eutrophic, otherwise the excess production of organic matter leads the system into disequilibrium and results in a change in water type.

Considering the whole biosphere, it clearly has been shown that detritus food chains, rather than grazing food chains, are the major pathways of energy flow. This is due largely to the stable trophodynamic nature of a detritus food chain. This stable relationship is maintained with a low but constant transfer rate of biomass from lower trophic levels to higher levels; this is possible partly because microorganisms in lower trophic levels have higher species diversity, and it has been shown that communities with high species diversity generally can adapt remarkably well to environmental changes. This characteristic of the detritus food chain is in contrast to that of the grazing food chain, where the trophodynamic relationship is maintained with a high transfer rate, resulting in a very unstable relationship.

V. ENERGY FLOW IN COMMUNITIES

The sun is the energy source upon which all life in the biosphere depends. At present, the energy of solar radiation can enter the biological cycle only through photosynthetic production.

The first step of carbon dioxide reduction is the chemical reaction forming 3-phosphoglyceric acid (PGA), which contains the carbon atom from carbon dioxide. PGA constitutes the first stable intermediate in the carbon dioxide reduction process (C_3 pathway or Calvin-Benson pathway).

Although the photosynthetic process takes place preponderantly in the euphotic zone, organotrophic as well as chemoautotrophic bacteria reduce carbon dioxide to form bacterial biomass. These processes cannot be neglected and are especially important in the extensive aphotic zone, which encompasses more than 95% of the sea by volume. At these greater depths, carbon dioxide seems to be reduced or fixed mainly by chemoautotrophic and organotrophic bacteria.[40] Both kinds of bacteria are known to occur in the hydrosphere, including chemoautotrophs which oxidize ammonium, molecular hydrogen, methane, or hydrogen sulfide as energy sources. On the other hand, Wood and Werkman[41] were the first to show by the isotopic tracer technique that carbon dioxide plays an active role in the metabolism of heterotrophs. Actually, all forms of life assimilate carbon dioxide, and the assimilation is an essential physiological function providing for the synthesis of indispensable metabolic intermediates. Among heterotrophs, many species of bacteria, yeasts, fungi, protozoans, and animal tissues have been shown to assimilate carbon dioxide. Life could not be maintained in the absence of available carbon dioxide. The fixation of carbon dioxide is made possible by the enzymes pyruvate carboxylase and phosphoenolpyruvate carboxylase, both of which lead to the synthesis of oxalacetate as the first stable product of carbon dioxide. Oxalacetate then enters the tricarboxylic acid cycle or is converted to carbohydrates through 3-phosphoglyceric acid. Accordingly, this route of carbon dioxide fixation is called the C_4-dicarboxylic acid pathway. All heterotrophic organisms can obtain some of their carbon for cellular biosynthesis from carbon dioxide using the C_4 pathway with the consumption of ATP and with reducing power ($NADPH_2$) generated either during the light reactions of photosynthesis or during oxidation of organic compounds. In this case, the organisms liberate more carbon dioxide in energy-yielding reactions than the amount fixed. The dark assimilation of carbon dioxide, through either chemosynthetic or organotrophic processes, may be considerable in coastal areas during certain periods of the year.[42] Carbon dioxide fixation also has been examined on the deep-sea floor. Bottom water and sediment containing a mixed bacterial population were collected from the Japan Trench (9500 m) and examined by Seki and ZoBell[40] for ^{14}C-carbon dioxide uptake. Bacterial carbon dioxide uptake in the sediment and bottom water during incubation at 1 atm ranged from 0.70 to 9.4 μg C per kg sediment per day and from 0.22 to 0.59 μg C per l water per day, respectively. These values possibly included participation of chemoautotrophic processes as well as organotrophic processes such as the C_4 pathway; if the experiments were performed *in situ*, the values would probably be 10 to 100 times lower due to the effect of hydrostatic pressure.

Organic matter is formed autotrophically through photosynthesis by plants and also through chemosynthesis by a certain group of bacteria. These bacteria can satisfy their energy requirements by utilizing simple inorganic compounds. All of the known organisms which comprise the chemosynthetic group are bacteria. In this process, ATP is generated by electron-transport phosphorylation during the oxidation of inorganic compounds serving as energy sources. The energy yields from the oxidation of various inorganic energy sources are as follows:

Species	Reaction	Element	$-\Delta G$ (kcal)
Thiobacillus	$H_2S + 1/2O_2 \rightarrow H_2O + S$	S	48.7
Thiobacillus	$S + 3/2O_2 + H_2O \rightarrow H_2SO_4$	S	120.3
Nitrosomonas	$NH_4OH + 3/2O_2 \rightarrow HNO_2 + 2H_2O$	N	65.2
Nitrobacter	$HNO_2 + 1/2O_2 \rightarrow HNO_3$	N	18.0
Hydrogenomonas	$H_2 + 1/2O_2 \rightarrow H_2O$	H	56.6
Leptothrix	$FeCO_3 + 1/4O_2 + 3/2H_2O \rightarrow Fe(OH)_3 + CO_2$	Fe	17

As these reactions provide energy to yield ATP, the organisms using these reactions are able to synthesize cell substance. When chemolithotrophs grow with carbon dioxide as the sole carbon source, they also must produce reducing power in addition to ATP. Although $NADPH_2$ can be produced directly by reduction of NADP, this production is coupled to many oxidation reactions when the redox potential of an energy source is lower than that of $NADPH_2$. H_2S or H_2 are well known energy sources in this process. When the redox potential of an energy source is higher than that of $NADPH_2$, then $NADPH_2$ must be generated with the consumption of ATP by a process called reversed electron transport.

In aquatic environments, chemolithotrophic bacteria are believed to fix approximately 1.5×10^5 tons of carbon annually. Thus their contribution is several orders of magnitude less than that of primary production by phytoplankton.[43]

From an energetics point of view, there is an important difference between photosynthesis and chemosynthesis. The source of energy for bacterial chemosynthesis in water may be from oxidation of methane, hydrogen, hydrogen sulfide, ammonia, or ferrous oxide. As these substances are derived from the decay of organic matter which originally is formed primarily through photosynthesis, chemosynthesis cannot be considered to be a major process in primary production in most locations of the biosphere, except for some unique locations such as submarine volcanoes or hot springs where hydrogen sulfide or other energy sources are supplied nonbiogenically for chemosynthesis. Therefore, chemosynthesis is a secondary process based on energy derived from organic matter created by photosynthesis, and photosynthesis is the only source of primary production. However, chemosynthesis usually involves carbon dioxide fixation and the primary formation of new biomass. Thus chemosynthesis may be considered as a type of primary production on the basis of its trophic position in aquatic food webs.

Particulate matter in the marine environment occurs in a great variety of shapes and in an almost continuous range of sizes. The particle distribution of marine communities should be within the range of 2×10^{-13} g (bacteria) and 1.3×10^8 g (whales). The conversion of particle volume to a sphere of equal volume allows all data on particle volumes to be normalized into a continuous size spectrum based on a logarithmic progression of particle diameters (Figure 17).

Sheldon et al.[44] showed that approximately equal concentrations of material occur at all particle sizes within the range from 1 μ to about 10^6 μ; i.e., for sizes from bacteria to whales in both surface and deep waters of the Antarctic or from bacteria to tuna in the equatorial Pacific Ocean. Some decrease of standing stock may occur as particle size increases, but the decrease from bacteria to whales is no more than a factor of two or four and is certainly less than one order of magnitude (Figure 18). The size range from bacteria to whales covers the whole of the marine food chain. Concentrations in the phytoplankton size range should be reduced by a factor of at least two to estimate living materials only.[44] The zooplankton concentrations are approximately correct as it is unlikely that many nonliving particles are present in that size range. The estimated concentrations of tuna and whales are believed to represent minimums for particles of those sizes.[44] The particle concentration in the bacterial size range should be reduced by a factor of at least two to estimate only living material, because the concentration has been determined by direct microscopy.[46] When these corrections are made to estimate standing stocks of living particles, the broken lines at constant concentration in Figure 18 represent the possible spectrum of living particles in the Antarctic and equatorial Pacific Oceans. The pattern of standing stocks are similar in each of the two oceans, although the absolute values differ by approximately a factor of ten. This could be expected because of differences in productivity of each region. This tendency for roughly similar amounts of particulate material to be present in any size range has ecological implications of considerable significance.

In a generalized aquatic food chain, relatively large predators feed on relatively small prey. The pattern of standing stock described above can be maintained only if the rate of particle production varies inversely with particle size. This is evident in Figure 19, which shows the relationship between production rates and particle sizes of marine organisms. A characteristic of bacteria and other nanoplankters is their rapid rate of reproduction, accomplished by vegetative cell division. Their biomass formation can be extremely rapid since the increase is by geometric progression. The rate of division of bacteria may be as frequent as once every few hours in the marine environment. The growth rate and size of a multicellular organism vary significantly during its lifetime, whereas a single-celled organism is less variable in these respects. Growth rate varies with temperature, but this effect is small relative to the scale in Figure 19. Further, quantitative relationships in each food chain seem to depend primarily on subtle interactions between the growth rates and metabolic efficiencies of predators and prey. For example, in a food chain including two well-known predator-prey links, *Clupea* to *Calanus* and *Calanus* to diatoms, production rates vary by roughly one order of magnitude for each step. Thus the standing stocks must be similar if the ecological efficiency is about 10% in the food chain.

In aquatic ecosystems, therefore, the standing stock of the microscopic primary producers has essentially reached a steady-state equilibrium, with both continual energy supply from solar radiation and continual energy removal by grazing of herbivores. Thus oscillation in the steady-state equilibrium of the lowest trophic level causes a delayed and amplified secondary oscillation of the predators. Accordingly, any trophic level in each ecosystem is led toward a steady-state equilibrium in which the degree of balance depends upon how far that trophic level is removed from the primary producers. However complex the equilibrium may be, the steady-state oscillation is in balance with the primary oscillation of primary producers.

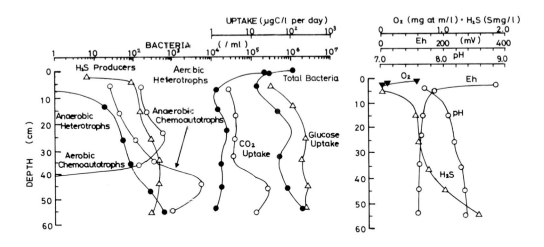

FIGURE 1.   Stratification of different microbial flora in a large water column of natural seawater from the intermediate water mass of Saanich Inlet, B. C., Canada. (From Seki, H., *Biological Oceanography of the North Pacific Ocean,* Takanouti, A. Y., Ed., Idemitsu Shoten, Tokyo, 1972, 487. With permission.)

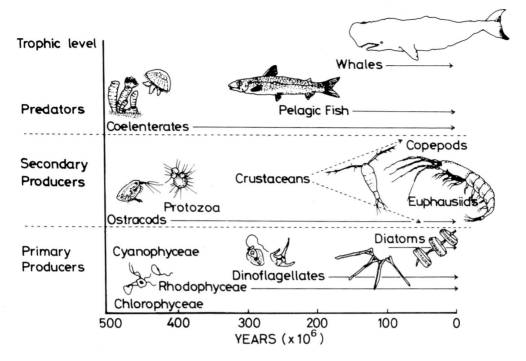

FIGURE 2.   The evolution of three food chains in the pelagic environment of the sea from the Cambrian to the present. (From Parsons, T. R., *S. Afr. J. Sci.,* 75, 536, 1979. With permission.)

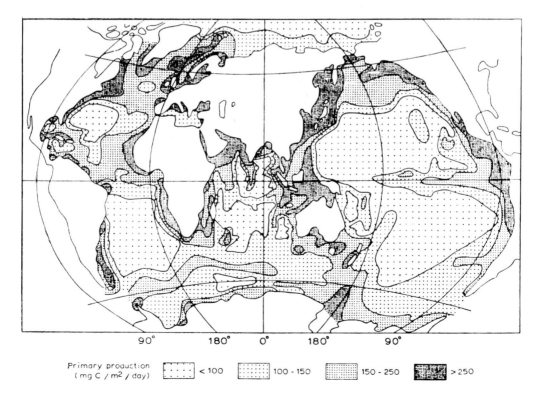

FIGURE 3.   Distribution of primary production in the world ocean. (From Parsons, T. R. and Takahashi, M., *Biological Oceanographic Progresses,* Pergamon Press, Oxford, 1973, 186. With permission.)

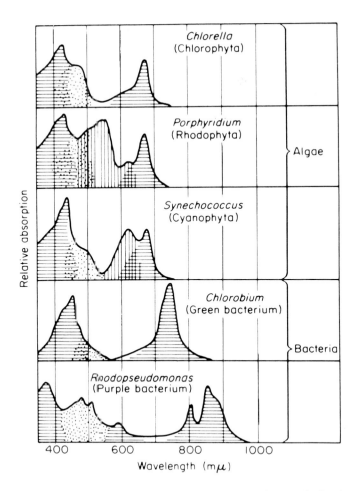

FIGURE 4. Absorption spectra of representative photosynthetic microorganisms. The contributions of the various classes of photosynthetic pigments are approximated as follows — horizontal lines: chlorophylls; stippling: carotenoids; vertical lines: phycobilins. (From Stanier, R. Y. and Cohen-Bazire, G., *Microbial Ecology,* Cambridge University Press, London, 1957, 56. With permission.)

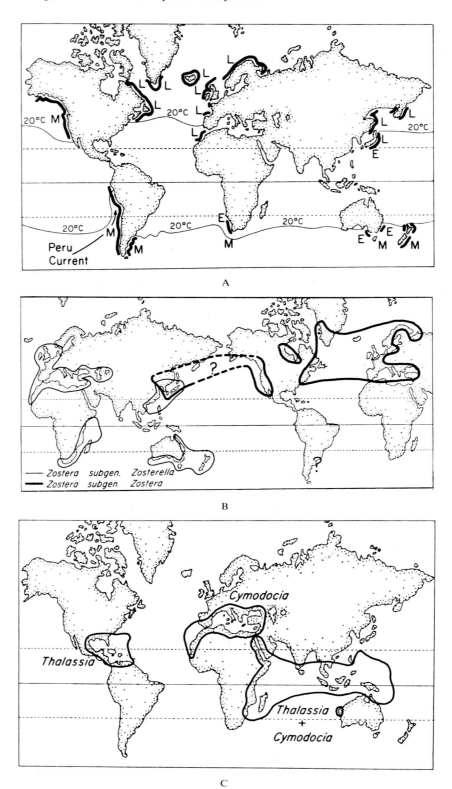

FIGURE 5.   The distribution of (A) *Laminaria* (L), *Macrocystis* (M), *Ecklonia* (E)., (B) *Zostera.*, (C) *Thalassia* and *Cymodocia.* (From Mann, K. H., *Mem. Ist. Ital. Idrobiol.*, 29 (Suppl.), 353, 1972. With permission.)

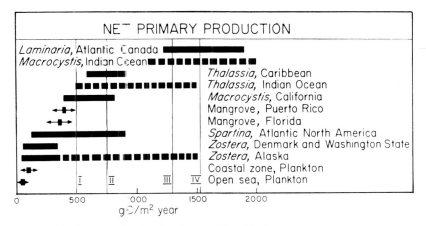

FIGURE 6. The range of net annual primary production of the major marine macrophyte systems, compared with phytoplankton. Broken lines indicate extrapolation from short-term data. (From Mann, K. H., *Mem. Ist. Ital. Idrobiol.*, 29 (Suppl.), 353, 1972. With permission.)

1. ARCTIC

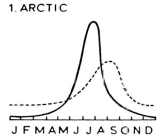

2. NORTH ATLANTIC

3. NORTH PACIFIC

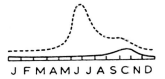

4. TROPICAL

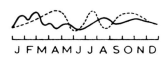

FIGURE 7. Seasonal cycles in plankton communities (———, changes in phytoplankton biomass; ———, changes in zooplankton biomass). (From Parsons, T. R. and Takahashi, M., *Biological Oceanographic Progresses*, Pergamon Press, Oxford, 1973, 186. With permission.)

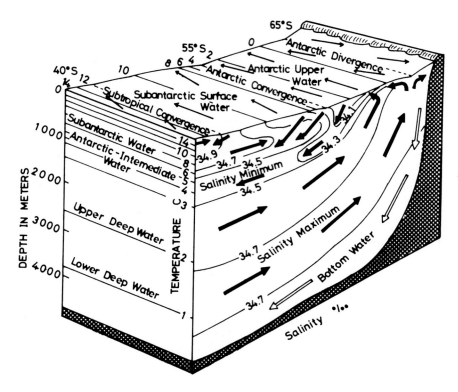

FIGURE 8.   Schematic diagram of the meridional and zonal flow in the Southern Ocean. The diagram represents the summer conditions; average positions of convergence and divergence shown. The upper deep water is best developed in the Atlantic sector. The south-going component in the lower deep water is weak or reversed in the Pacific. (From Knox, G. A., *Antarctic Ecology,* Holdgate, M. W., Ed., Academic Press, London, 1970, 69. With permission.)

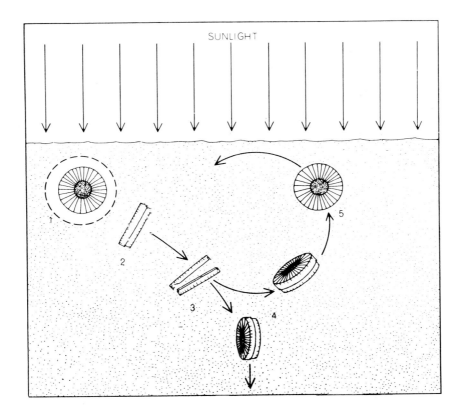

FIGURE 9. Phytoplankton cell is slightly denser than seawater and under absolutely quiet conditions would slowly sink to the bottom. In this way the cell can move from a small parcel of water (broken circle) from which it has removed all the available nutrients (black dots) into a parcel still containing these substances. As the cell sinks it divides, and losses from the population in the surface waters that constitute the euphotic zone are continually made good by upward turbulence, which returns some of the products of cell division to the surface layer. The particular phytoplankton shown is a diatom of the genus *Coscinodiscus*. (From Hutchinson, G. E., *The Biosphere*, W. H. Freeman, San Francisco, 1970, 3. With permission.)

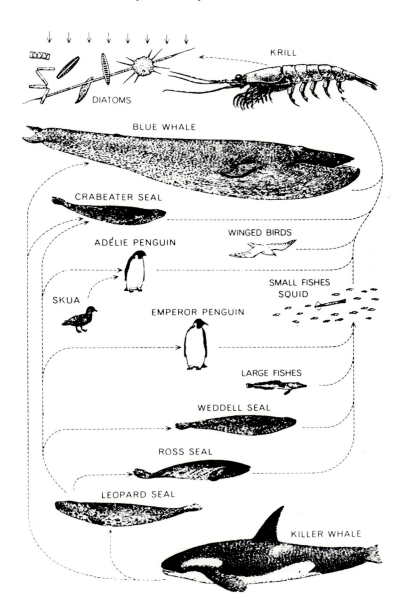

FIGURE 10.   The food web in the Antarctic Ocean. Diatoms utilize the energy of the sun (short arrows) to transform nutrients in the seawater into living tissue. The diatoms are fed on by the krill *Euphausia superba.* The krill is in turn the food of whales, penguins and other birds, the crabeater seal, squid, and fishes. In addition to this basic food chain, there is predation by seals, penguins, and large fishes on squid and small fishes; the leopard seal preys on penguins and other seals and the skua eats penguin eggs and chicks. (From Murphy, R. C., *Sci. Am.*, 207, 186, 1962. With permission.)

FIGURE 11.   Simplified food chain shows how each step in the process involves a diminishing return. That is, it takes 100 units of phytoplankton, such as diatoms, to grow ten units of krill, which in turn is enough to grow only one unit of its predator, the whale. (From Murphy, R. C., *Sci. Am.*, 207, 185, 1962. With permission.)

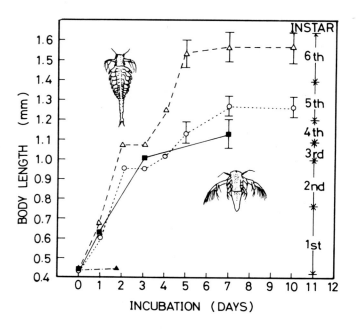

FIGURE 12. Growth of *Artemia* on bacteria. (Bacteria are grown on: Δ, 0.01% Bacto-peptone; O, 0.005% Bacto-peptone; ■, water extract of sediment equivalent to carbon content in 0.005% Bacto-peptone; ▲, no bacteria or medium added. Vertical bars indicate the range of values obtained from ten measurements.) (From Seki, H., Skelding, J., and Parsons, T. R., *Limnol. Oceanogr.*, 13, 440, 1968. With permission.)

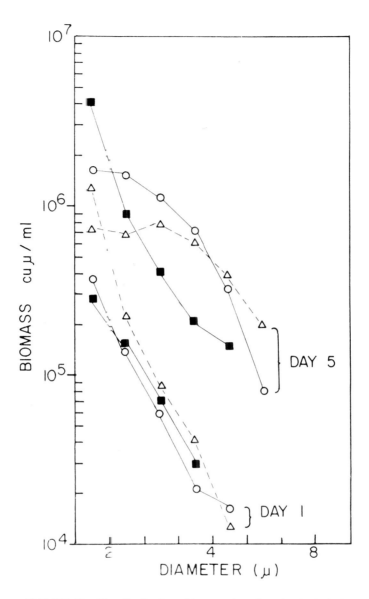

FIGURE 13. Size distribution of two species of marine bacteria incubated in an enriched seawater medium for 5 days (■—■ *Pseudomonas* sp.; Δ—Δ *Chromobacterium* sp.; ○—○ mixture of both species). (From Parsons, T. R. and Seki, H., *Organic Matter in Natural Waters*, Hood, D. W., Ed., University of Alaska, Fairbanks, 1970, 1. With permission.)

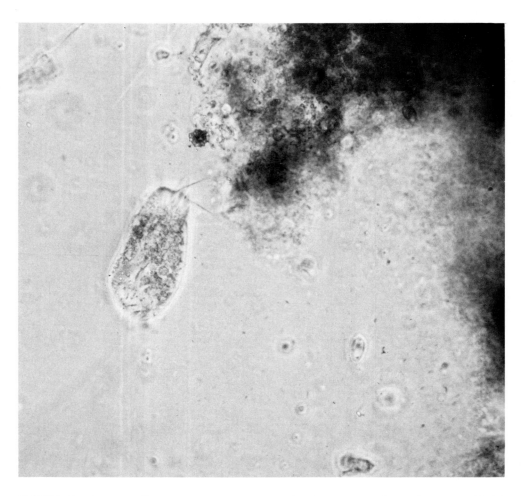

FIGURE 14.   Bacteria, ciliate, and vorticella in the matrix of an aggregate from a hypereutrophic lake, Lake Kasumigaura in Japan.

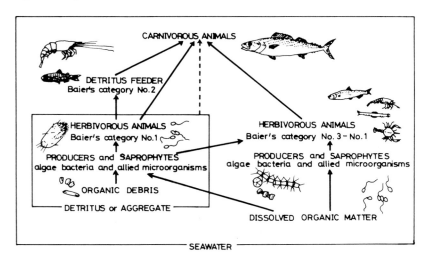

FIGURE 15.   Scheme of marine food web, with particular reference to aggregates in seawater. (From Seki, H., *Mem. Ist. Ital. Idrobiol.*, 29 (Suppl.), 245, 1972. With permission.)

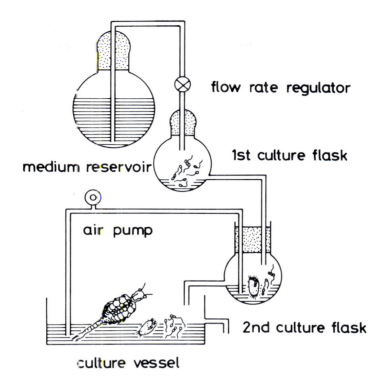

flow rate regulator

medium reservoir

1st culture flask

air pump

2nd culture flask

culture vessel

FIGURE 16. A continuous culture system for the culture of bacteria, protozoans and brine shrimps. Bacteria and protozoans were isolated from the sea water of Aburatsubo Inlet, Japan. (From Seki, H., *J. Oceanogr. Soc. Jpn.*, 22, 105, 1966. With permission.)

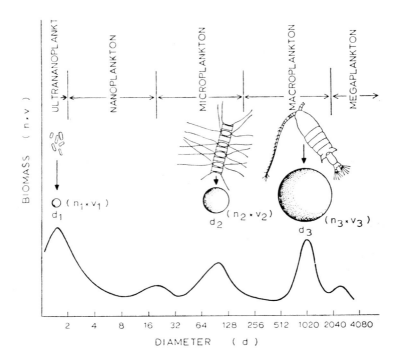

FIGURE 17.   Particle spectrum representing biomass (n × v) of material in different size categories determined by the diameter (d) of a sphere equivalent in volume (v) to the original particle times the number of particles (n). (From Parsons, T. R. and Takahashi, M., *Biological Oceanographic Progresses,* Pergamon Press, Oxford, 1973, 186. With permission.)

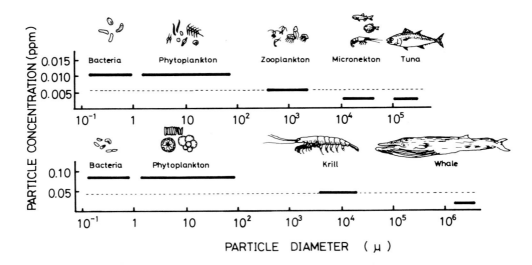

FIGURE 18.   Estimates of standing stock (thick lines) in the oceans; above — equatorial Pacific, below — Antarctic. The thin broken line is an estimate of the true or potential standing stock of living material. (From Sheldon, R. W., Prakash, A., and Sutcliffe, W. H., Jr., *Limnol. Oceanogr.*, 17, 327, 1972. With permission.)

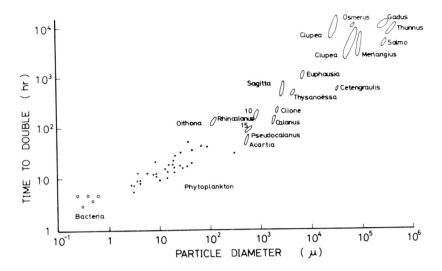

FIGURE 19. The relationship between production rate and particle size. The numbers near to the *Rhincalanus* patches indicate the temperature at which the growth took place. The uppermost of the two *Clupea* areas represents *C. sprattus*. The lower area represents both the Atlantic (*C. harengus*) and Pacific (*C. pallasi*) herring. (From Sheldon, R. W., Prakash, A.. and Sutcliffe, W. H., Jr., *Limnol. Oceanogr.*, 17, 327, 1972. With permission.)

Table 1
SOLAR ENERGY AVAILABLE FOR PHOTOSYNTHESIS IN THE OCEAN, OCEAN PRODUCTIVITY, AND OCEAN CROPPING[12]

Solar Energy Available for Photosynthesis in J/Year

1. Intercepted by Earth	5.0×10^{24}
2. 40% reaches surface	2.0×10^{24}
3. 50% of (2) is infrared	1.0×10^{24}
4. 40% of (3) is reflected	6.0×10^{23}
5. Available in ocean, 75% of (4)	4.0×10^{23}
6. Assuming 2% efficiency in PS	8.0×10^{21}
7. Needed per g of C assimilated	5.0×10^{4}
8. Maximum assimilation per year	1.6×10^{11} tons of C

Ocean Productivity

Carbon assimilation per year	100—200 g C/m²
Area of oceans	3.5×10^{14} m²
Annual productivity	5.3×10^{10} tons C
Maximum productivity	1.6×10^{11} tons C

Ocean Cropping

Commercial fish catch for 1961	$\sim 10^{11}$ lbs. fresh weight (FAO Fishery Statistics)
	$\sim 5 \times 10^{7}$ tons fresh weight
In terms of carbon	$\sim 3.8 \times 10^{6}$ tons
Annual productivity	5.8×10^{10} tons
Maximum productivity	1.6×10^{11} tons

Chapter 2*

FUNCTIONS OF AQUATIC ENVIRONMENTS

I. INTRODUCTION

Approximately 72% of the surface of the Earth is covered with water, mostly salt water (Figure 1). More than 98% of the water is present in the sea or its sediments.[47] Less than 2.0% of the water occurs in freshwater lakes, rivers, ponds, reservoirs, ground waters, and atmospheric moisture. The Vitiaz Deep (11° 20.9′ N, 2° 11.5′ E), found in 1957 in the Mariana Trench, has a depth of 11,034 m and is the deepest known part of the ocean. Due to its great depth and volume, the marine environment is estimated to provide approximately 300 times the inhabitable space provided by the terrestrial and freshwater environments. In fact, the marine habitat provides livable space for at least some form of life from the surface waters to the deepest parts of the ocean, whereas inhabitable areas of the terrestrial environment are restricted mainly to the immediate surface and penetrate to a maximum depth of 1 m.[48] Superficially, seawater in the oceans seems to form a continuous, homogeneous body of water without boundaries but divided into the five major regions of the Arctic, Antarctic, Atlantic, Pacific, and Indian Oceans. Actually, however, many water masses are formed at the sea surface and sink and spread in a manner that depends on their density relative to the general distribution of density in the ocean. Some elements in a large water mass tend to be unchanged, such as the average distribution of temperature, salinity, and oxygen. These conservative elements are not seriously altered by processes of diffusion and advection. However, external processes tend to modify these conservative elements at the boundary surface of a water mass. Each water mass is further characterized by stratification and circulation (Figure 2). Vertical circulation may extend to a great depth; greater or smaller water masses flow continuously in systems extending in diverse directions over vast areas, irrespective of latitude. The vertical stratification of several water masses is detectable in the oceans; i.e., surface, upper, intermediate, deep, and bottom waters. Both the general oceanic circulation and the distribution and stratification of water masses are driven by the energy of solar radiation. Physical processes taking place at the boundary between the water and the atmosphere are either directly or indirectly responsible for the behavior of the oceans. The general atmospheric circulation is responsible for the establishment of the major surface currents of the oceans; the wind system near the sea surface is particularly responsible for this phenomenon. Not only does the local wind field affect ocean currents, but so also do changes of the general atmospheric circulation over large oceanic regions. The thermal structure of the upper strata in the oceans is governed by the main thermocline. The thickness of the nearly homothermal water layer above the thermocline depends not only on the absorption of solar radiation, heating and vertical mixing, but largely on the type of circulation in the top layers. The main thermocline gradually weakens with increasing latitude and finally disappears in regions poleward of the subtropical convergences. It also weakens in the immediate vicinity of the equator.

In the Pacific Ocean, the depth of the main thermocline changes from less than 50 m in the east to approximately 200 m in the west. In the Atlantic Ocean, that depth varies between about 20 and 250 m with a general shallowing toward the equator and toward the east.

The deep water mass is the largest and occurs between depths of approximately 1000

* All figures and tables for Chapter 2 appear after the text.

and 4000 m. The deep water of the Atlantic Ocean is the most clearly differentiated. It is governed by the intrusion of high salinity water from the eastern Mediterranean Sea, an area where evaporation exceeds precipitation. As a result of its high salinity of 39.1%, it sinks from the surface into depths of about 500 m while spreading westward. The intrusive effect of Red Sea water on the stratification of Indian Ocean deep water is believed to be very similar to that of the Mediterranean water in the Atlantic Ocean. In the Pacific Ocean, an intrusive source of high salinity into the deep water is absent, and therefore no outstanding indicators are available to differentiate the deep sea stratification (Figure 3).

The bottom water of the oceans is formed in polar regions, especially in waters of the southern hemisphere close to the Antarctic continent. A major source of Antarctic bottom water is known in the Weddell Sea.

Processes that increase the density of surface water, such as cooling, evaporation, and ice formation during the winter months, create vertical instabilities of the water that make convection processes possible between the sea surface and the bottom. In balance with the sinking of water at regions of convergences to form deep water masses, water ascends at areas of divergences. Upwelling is particularly prevalent along the western coasts of continents, where prevailing winds carry surface waters away from the coasts. This upwelling brings nutrient-rich water of greater density and lower temperature toward the surface.

II. TEMPERATURE

Temperature is the most important environmental factor affecting the equilibria and rates of chemical reactions, growth rates, nutritional requirements, and the enzymatic and chemical composition of cells. All biochemical reactions require appreciable temperature in order to be brought beyond the threshold at which biological processes occur from a state of nearly no kinetic energy of matter.

In 1889, Arrhenius studied the effect of temperature on the rate of sucrose hydrolysis and formulated a theory that can be expressed as

$$\log_{10} v = \frac{-\Delta H^{\ddagger}}{2.303\ RT} + C \quad \text{(The Arrhenius equation)}$$

where v represents the velocity of the reaction, R is the gas constant (1.98646 cal/°C/mol), T is the absolute temperature, A is a constant, and $\Delta H^{\ddagger}$ is an energy term referred to as the activation energy.

Following Arrhenius, extensive works have shown that the rate of biological processes also is described with considerable accuracy by the Arrhenius equation, provided that the range in temperature is not too large. Crozier and his colleagues (1924 to 1926) studied the temperature effect on a large number of various biological processes and established that there is a considerable temperature range over which the temperature-activity relation closely follows the Arrhenius equation. Each biological process has a definite temperature characteristic. The apparent temperature characteristic of an overall process, such as bacterial respiration or bacterial growth, is defined by the temperature characteristic of the master reaction. Over a wide temperature range, the overall process usually is controlled by more than one master reaction. Thus the Arrhenius equation also can be applied to biological reactions, including complicated physiological processes in a limited temperature range.

The upper temperature limit of life is set by the stability of the indispensable molecules making up organisms. A given temperature will lead to change of a known key molecule and thus kill a given cell. Since hydrogen bonds are weak in general, these bonds are broken during the heat denaturation of proteins, and losses in biological activity result.

On the other hand, temperatures below the minimum growth temperature of an organism are believed to inhibit enzyme function and the initiation of protein synthesis, because many enzymes are unusually sensitive to feedback inhibition at low temperature. Thus the response of a particular organism to temperature can be described by the use of the cardinal temperatures; i.e., the minimum growth temperature, the optimum growth temperature and the maximum growth temperature.

The minimum growth temperature is the minimum temperature at which an organism can continue to grow. The optimum growth temperature is the temperature at which the organism grows best. The maximum growth temperature is defined as the temperature above which growth is not possible. The optimum temperature is always nearer the maximum than the minimum.

As defined by these cardinal temperatures, it is possible to characterize three groups of organisms: psychrophiles with low temperature optima; mesophiles with midrange temperature optima; and thermophiles with high temperature optima. The term "psychrophile" refers to a cold-loving organism; i.e., a true or strict psychrophile is an organism with an optimal growth temperature of 15°C or lower and a minimal growth temperature of 0°C or lower. Even in many frozen materials, there are usually microscopic pockets of liquid water where such microorganisms can grow.

As more than 90% by volume of the ocean is 5°C or less, the oceans should be a desirable natural habitat for psychrophilic organisms.[51] Because of this situation, it is relatively easy to isolate psychrophilic bacteria from waters below the oceanic thermocline as well as from Arctic and Antarctic waters. Bacteria with the ability to grow at the low temperatures of the Antarctic Ocean have been known to exist since the early investigations by Ekelöf[52] during the Swedish Antarctic Expedition from 1901 to 1903. A bacterium recently isolated by Morita[51] has an extremely low temperature optimum of 4°C and a maximum growth temperature of 10°C (Figure 4). Organisms also are known which grow at 0°C but have optima of 25 to 30°C; these have often been called facultative psychrophiles.

Organisms with optima between 25 to 40°C are called mesophiles. *Escherichia coli* is a well-known example of a mesophile from man, having an optimum growth temperature of approximately 39°C. The most common microbial environments with temperatures over 30°C are the bodies of birds and mammals.

Organisms that grow at temperatures above 45 to 50°C are called thermophiles. Temperatures as high as these are found only in certain restricted natural areas such as hot springs.

The temperature range throughout which life can exist is from below freezing up to the boiling point of water. However, no single organism can grow over this whole temperature range, and the usual range for a given organism is approximately 30 to 40°C.

Mesophiles grow at higher temperatures than do psychrophiles. The growth rates of mesophiles at their optimum temperatures are faster than those of psychrophiles because rates of enzymatic and chemical reactions increase at higher temperature. For the same reason, thermophiles are able to grow faster than mesophiles at their respective optimal temperatures (Figure 5).

The temperature of the aquatic environment ranges from −1.9°C to 40°C, and is usually less than 30°C, depending largely on latitude and season. The freezing point of seawater having a salinity of 35% is −1.9°C. The seasonal variation of ocean surface temperature is generally less than 10°C, being least in tropical waters and greatest in temperate zones; the change in sea surface temperature closely follows the annual variation of incoming solar radiation. Higher temperatures occur only in limited localized areas of the marine and freshwater environments. More than 90% of the marine environment, by volume, is colder than 5°C, and the thermal distribution of inland waters depends primarily on their geographical positions and depths. Strict psychrophiles may be found in an environment such as the deep water of the oceans, but facultative psychrophiles may be much more widely distributed in the upper layers of the hydrosphere where biological processes are highly active compared to those of the deeper layers. For example, in the Atlantic Ocean areas, surface temperatures of less than 10°C cover only about 28% of the total area, and temperatures of less than 5°C cover only about 18%; this surface temperature range should be favorable for the activity of facultative psychrophiles.

Even in the ocean, extreme high-temperature environments are found in association with volcanic phenomena. In the deep-sea environment (2800 m) of the Galapagos hot spring area, thermophiles exist and have a good rate of metabolism.[54]

As microbial activity has little effect on the temperature of localized environments by biochemical heating (amounting to only a few degrees Celsius), psychrophiles can manage well in cold environments by having a higher content of unsaturated fatty acids in the cell membrane than do other organisms. As a consequence of this property, the membranes of psychrophiles remain semifluid at low temperatures, and the active transport systems for nutrient uptake function well at low temperatures.

A change in the maximum growth temperature of a typical marine psychrophilic bacterium, *Vibrio marinus* MP-1, was shown to be dependent on the various ions present in the culture medium (Figure 6). The order of effectiveness in restoring the normal growth temperature was Na^+, Li^+, Mg^{2+}, K^+, Rb^+, and NH_4^+ when each of these ions was added to dilute seawater. In a completely defined medium, the highest maximal growth temperature was 20.0°C at 0.40 mol NaCl. A decrease in the maximum growth temperature was observed at both low and high concentrations of NaCl.

A temperature-salinity relationship was shown for the induction of glutamic dehydrogenase in *V. marinus* MP-1. This interaction of salinity and temperature may be important in the near-shore environment, because it will determine whether or not a microorganism will have the ability to synthesize inducible enzymes to cope with any foreign substrates that might enter the water.[51]

The ability of this barophobic psychrophilic bacterium to synthesize macromolecules, such as protein, RNA, and DNA, was not affected by hydrostatic pressures up to 200 atm. Protein and RNA, but not DNA synthesis was adversely affected at 400 atm whereas at 500 atm there was a rather sharp decline in all synthetic rates. At 1000 atm virtually no synthesis occurred.[51]

Pressure also is found to inhibit the transport systems for nutrient assimilation of those marine bacteria with the ability to grow at low temperature.[51] Once amino acids are transported into their cells, the respiratory mechanism needed to produce carbon dioxide from the amino acids does not appear to be inhibited. Immediate pressurization of their cells in the presence of an amino acid brings about inhibition of transport, and this inhibition is reversible when the pressure is released.

Most aquatic microorganisms are thermosensitive. Their enzymes are very sensitive to higher temperature and are rapidly inactivated at temperatures of 30 to 40°C. This is a common characteristic of psychrophiles, or all organisms having the ability to grow at low temperatures.

In the temperate inshore waters of Narragansett Bay in the western Atlantic Ocean, water temperature ranges vary from −2 to 23°C. Such a marked seasonal variation influences the temperature characteristic of microbial growth. The dominant microflora occupied a 12 to 15°C growth range throughout the year. This growth range shifted 12 to 15°C according to the seasonal variation in water temperature. The water temperature curve and the growth range curve of microorganisms were out of phase by approximately 2 months (Figure 7), and both were sigmoidal. Therefore the minimum, optimum, and maximum growth temperatures of the dominant microbial communities change continuously as a result of continual changes in water temperature. There is, however, a marked lag between the water temperature and the thermal response of natural microbial communities.

The maximum growth temperature of the dominant microbial communities is approximately 10°C higher than the *in situ* temperature. This must be a natural mechanism to protect a certain thermal group from sporadic or short-term impacts of heat shock. Similarly, organisms living in estuarine and coastal regions, where thermal variation is appreciable, are more eurythermal than those inhabiting tropical or arctic surface waters or deep waters where the thermal condition is almost constant throughout the year. Incidentally, procaryotes are more eurythermal than eucaryotes; nonphotosynthetic organisms are more eurythermal than photosynthetic organisms; and structurally simple organisms are more eurythermal than more complex ones.

One of the most convincing hypotheses of microbial evolution is that thermophiles were the first to evolve, followed by mesophiles and then psychrophiles. The evolution of psychrophilic bacteria is due to many genetic events taking place in the predominantly cold environment of the biosphere.[56]

III. SALINITY

Organisms can be ecologically classified on the basis of their salt requirement: nonhalophiles that grow best in media containing less than 2% salt, slight halophiles that grow best in media containing 2 to 5% salt, moderate halophiles that grow best in media containing 5 to 20% salt, and extreme halophiles that grow best in media containing 20 to 30% salt.[57] Slight halophiles, moderate halophiles, and extreme halophiles are collectively called halophiles; they are found in habitats where salt is a component of the environment.

Extreme halophiles occur in great numbers in the Great Salt Lake, the Dead Sea, and other natural waters of extremely high salinity. These extremely halophilic organisms are exclusively microorganisms. They are widespread in nature, not only in aquatic environments but in any location. Such halophilic bacteria are red in color and are commonly introduced into food materials which are preserved with solar salt. When they are introduced into salt fish, the fish become red and are called pink or pinkeye. This reddening is due to the multiplication of bacteria belonging to the *Halobacterium* group or the *Sarcina-Micrococcus* group. Extremely halophilic organisms are mostly bacteria, but some are green algae, such as *Dunaliella salina* which lives in the vegetative state at salt concentrations near saturation.

Moderate halophiles frequently occur in large numbers in curing brines. Most of the moderately halophilic bacteria belong to the genera *Nitrosomonas, Thiobacillus, Pseudomonas, Vibrio, Achromobacter, Bacteroides, Micrococcus, Sarcina, Pediococcus,* and *Bacillus*. They are colorless in contrast to the extreme halophiles, all of which are pigmented.

Typically, marine bacteria are slightly halophilic organisms. Most of them are Gram-negative, nonspore-forming, and have motile rods. They grow best in seawater media or in an isotonic mineral solution. Their growth usually is inhibited by salt concentrations of culture media lower than 1.5% or higher than 5%.

Among nonhalophilic organisms, the obligate anaerobic spore formers seem to be the most sensitive to salt. Many of them are completely inhibited at 5% sodium chloride. Gram-negative rods generally are completely inhibited at higher salt concentrations up to 15 or even 20%. This is because many Gram-negative bacteria are plasmolyzed when suspended in salt solution due to their plasma membranes being impermeable to salt. Gram-positive bacteria, on the other hand, generally cannot be plasmolyzed because their plasma membranes are permeable to salt.

In high concentrations of salt, enzymes may be precipitated or inactivated and thus the speeds of biochemical reactions are reduced sharply. Even in such cases where enzyme losses do not occur in the solution, the alteration of viscosity may modify reactions in which diffusion plays a part. Moreover, a high salt concentration will reduce the water activity and so retard biochemical reactions.

The requirement of halophilic organisms for salts in media has erroneously been believed to reflect a requirement of these organisms for media of suitable osmotic pressure to maintain cell integrity. However, many bacteria which require seawater in the medium for growth have been shown to have a specific requirement for at least minimal amounts of Na ions. Sodium is necessary for marine bacteria to transport metabolites into their cells and to maintain cytoplasmic membrane proteins in the proper conformation, thereby permitting tight packing of membrane subunits and preventing leakage of intracellular solutes. Sometimes, when salts are adsorbed by protein ions, the net charge of the protein ions is altered in the process. If the protein ion is an enzyme, this process alters the biochemical reaction between the enzyme and its substrate. Such modification of the charge on the enzyme and of its activity is usually not lethal, even for nonhalophilic microorganisms. Specific metal ions are required for the activation of many enzymes. Such ions as Na^+, K^+, Mg^{2+}, Ca^{2+}, and Zn^{2+} are common metal-ion cofactors. These metal ions are either integral parts of the protein structure of the enzymes or are transient associates of the enzymes. Sodium is required especially by marine microorganisms for growth, whereas closely related freshwater forms may be able to grow in the complete absence of sodium. The specific requirement for sodium cannot be satisfied by replacement, even with a chemically related ion such as potassium. Potassium ions also are required for the stability of the cell wall of *Halobacterium*, an extremely halophilic bacterium.

The mean salinity of freshwater lakes and rivers of the world has been determined to be approximately 100 ppm, but individual variations are enormous. The water composition of freshwater lakes tends to approximate that of rivers. In some cases, however, the composition and concentration are modified by the inflow of water from old salt beds, which increases the salinity to those high values present in the Great Salt Lake and the Dead Sea.

Since the salinity of water masses in the pelagic regions of the oceans ranges from 33 to 37%, this salt environment is most favorable for the growth and activities of slight halophiles. On the other hand, the salt environment in coastal regions of the oceans is highly dynamic. Major factors that determine the surface salinity are differences in evaporation minus precipitation, turbulent mixing, and advection by currents. The runoff of river water, ice melting, and freezing also figure locally. The typical estuary, where freshwater flows into the marine environment, usually is characterized by transitions in salinity ranging from 0 to approximately 35%. Low density freshwater flows over the denser saltwater for many kilometers seaward into the pelagic regions of the oceans.

The effect of salinity on microbial communities has been studied in the river and estuary of the Nanaimo River (Vancouver Island, Canada), where the relative heterotrophic potential of water samples was established over the salinity gradient of 0 to 25% throughout the year. Only the results obtained in July and November are shown in order to demonstrate the maximum changes from summer to winter (Figure 8). In river water samples with zero salinity, the maximum microbial heterotrophic activity was at 0% NaCl; this activity decreased sharply in both summer and winter samples, so that less than 1% of the heterotrophic activity of river water occurred above a salt concentration of 2%. In waters having salinities of 10 to 20%, the introduction of halophilic bacteria was particularly marked in November when there were two peaks of heterotrophic activity, one optimal at at 2% NaCl and the other optimal at 4%. In samples having a salinity of 25% the nonhalophilic bacteria contributed only a small percentage to the glucose uptake compared with the activity of the halophilic bacteria having optimal activity at 4% NaCl. In both the July and November samples the maximum microbial heterotrophic activity of seawater may occur at salt concentrations of 2% NaCl. This concentration would appear to be optimal for a mixed population of bacteria, consisting of both euryhaline nonhalophiles and slight halophiles. In addition to supporting the heterotrophic activity of a mixed population, the 2% NaCl sample might also represent an optimal mixture of growth factors from the land and inorganic nutrients from the sea. It is also quite apparent that the heterotrophic activity of samples collected in July was one or two orders of magnitude greater than the activity of samples collected in November. This is partly because the higher summer temperature accelerates the active transport system of nutrient uptake by bacteria. Also, the survival of freshwater bacteria in the marine environment is greatest during the summer, when the largest amount of soluble organic matter is present in seawater. This may be because seawater inhibits the growth of nonhalophilic river organisms unless trace amounts of organic chelating agents are present. Actually, the amount of allochthonous organic matter contributed annually by land was comparable to the autochthonous organic matter produced in the Strait of Georgia.

Microorganisms and their activities also were studied in the lower reaches of the River Teshio in Japan with special reference to the mixing of freshwater and seawater.[59] The nature and processes observed therein were similar to those studied in the Nanaimo River, and they may be generalized as common to nonglacial rivers in sub-temperate zones. The profile of microbial activity reflected the transition from freshwater to seawater in the River Teshio (Figure 9); i.e., the output of particulate organic materials through self-purification was extremely low in the lower reaches of the river, probably because of removal of freshwater microorganisms through their aggregation and precipitation during mixing of the freshwater and seawater. Such depression of microbial activity favors a greater supply of organic energy sources from the river to the sea, and also aggregated freshwater bacteria can be grazed easily by marine predatory protozoans.[24] In the lower reaches of rivers, therefore, the decrease in microbial activity must be due to a decreased activity of freshwater microorganisms (originally of mesotrophic type) and not to oligotrophic marine microorganisms.

In extreme cases, the salinity in isolated regions such as lagoons and the Red Sea may exceed 40% up to 100%. In these areas, euryhaline microorganisms may become predominant. Procaryotic microorganisms tend to be more euryhaline than eucaryotes, but no single procaryote species can grow in the whole range of salt concentrations which occur in the biosphere

IV. HYDROGEN-ION CONCENTRATION AND REDOX POTENTIAL

The pH of natural waters varies from as low as 1.7 in some volcanic lakes containing free sulfuric acid to 12 or more in some alkaline lakes rich in soda.[47] Most natural aquatic environments, however, have pH values between 5 and 9. The usual range for open lakes is between 6 and 9. The range encountered in the sea is between 7.5 and 8.4. Wherever seawater is in equilibrium with the carbon dioxide in the atmosphere, the pH is between 8.1 and 8.3. Higher pH values may occur when photosynthetic activity has reduced the content of carbon dioxide.

The pH is related to the cation, carbonate, bicarbonate, and CO_2 concentrations in natural waters, and a good approximation is given by the equation:

$$pH = 10.7 - \sqrt[3]{C_m} - \log \frac{HCO_3^-}{P_{CO_2}}$$

where C_m is the total concentration in equivalents per liter. Natural waters are sometimes slightly undersaturated with respect to atmospheric CO_2, presumably as the result of photosynthesis, but more often they are slightly supersaturated. In acid lakes, however, the low pH must be due to acids other than carbonic. Sulfuric acid is the most likely component in these lakes, although colored organic acids also may be important in certain lakes.

Natural waters contain a variety of substances, and these substances vary greatly both qualitatively and quantitatively. Natural waters differ not only in the level of hydrogen-ion concentration but also in the buffer effect. Waters with acid or base have greater buffering action, which prevents rapid changes in hydrogen-ion concentration because the initial ionization is low. When newly entering substances affect the hydrogen-ion concentration in any water, the remaining undissociated molecules become ionized until a new equilibrium is established which has about the same hydrogen-ion concentration as before. Waters containing large amounts of dissolved matter are most likely to show a high buffer effect, while a low buffer effect is to be expected in waters with very low quantities of dissolved material. In the absence of carbonates, additions or reductions of the free carbon dioxide would result in a respective increase or decrease of carbonic acid and would alter the hydrogen-ion concentration. In the presence of carbonates, however, the addition or withdrawal of carbon dioxide immediately results in the reestablishment of the origial equilibrium, and thereby the hydrogen-ion concentration tends to remain the same. This mechanism holds only within the inherent limits of the buffer substances.

The effect of inflowing acids or bases on the pH equilibrium of any natural water will depend on its buffering action. Soft waters have low buffering action, and sudden changes in hydrogen-ion concentration may sometimes occur in them following heavy rains or freshets. Hard waters, on the other hand, have high buffering action and a similar inflow may either have a small effect or no appreciable effect unless the entry of substances was great.

In localized environments, conditions also may sometimes become too acidic or too alkaline for optimal growth of aquatic organisms. Under such peculiar conditions, the pH of tide pools, small bays, and estuaries may sometimes exceed the values observed in open waters.

A bacterium, *Thiobacillus thiooxidans*, is unable to grow or to produce sulfuric acid by oxidizing sulfur until the pH drops to 0.7. An eucaryotic alga, *Cyanidium caldarium*, is able to grow at a pH near zero. In acid environments, these organisms can maintain a pH close to neutrality either by keeping H^+ ions from entering or by actively expelling H^+ ions as rapidly as they enter. The cell wall is believed to play some role in keeping hydrogen ions from entering. Neutrality inside the cells is necessary because there are many acid- and alkali-labile cellular components. Chlorophyll, DNA, and ATP are known to be destroyed by acid.

At the other extreme, the bacterium, *Bacillus circulans*, grows in media of pH 11.0. The blue-green alga, *Plectonema nostocorum*, grows on media of pH 13.0. In order to grow in extremely alkaline environments, these microorganisms must keep the pH inside the cells close to neutrality, because RNA and phospholipids are easily destroyed by alkali. The regulation of a neutral pH inside the cells can be accomplished partly by enzymatic mechanisms. For example, there is a group of amino acid decarboxylases with optimal activity at approximately pH 4.0 and almost no activity at pH 5.5; this causes a spontaneous adjustment of the pH towards neutrality from the acid range. Another group of amino acid deaminases have optimal activity at approximately pH 8.0; this causes a spontaneous adjustment of pH towards neutrality from the alkaline range. Therefore, microorganisms can adjust the cellular pH to close to neutrality although they are found in habitats over a wide pH range. Such adjustment is especially important biochemically, because the optimum pH for intracellular enzymes is usually around neutrality. On the other hand, enzymes in the periplasm and extracellular enzymes may have pH optima for activity near the pH of the environment.

Organisms themselves alter the pH values of their environments through their activities, either by release of metabolic products into the environment or by selective removal of substances from the environment. Microorganisms that ferment carbohydrates to produce acids may lower the pH of their environment, often by as much as two orders of magnitude; or the pH may be raised through their microbial actions, usually because ammonia is released by the deamination of amino acids or other nitrogen-containing organic compounds. On the other hand, organisms growing on ammonia reduce the pH of their environment when removing the ammonium ions, whereas those growing on nitrates raise the pH in removing nitrate ions.

The primary energy source for all organisms is chemical energy supplied by chemical substances. Using a chemical compound as an energy source always involves an oxidation-reduction reaction. These reactions in living systems are so essential that life is sometimes defined as being a continuous oxidation-reduction reaction.

The basis of an oxidation-reduction reaction is electron transfer, and many of these reactions in biological systems are accompanied by the transfer of hydrogen and oxygen. The electrical work in the passage of a Faraday of electricity (F) against a potential difference (E) is equal to the work done in transporting an electron; this is measured electrically, usually in reference to a hydrogen (H_2) electrode as a standard. Electrode potentials referred to this standard are measured in volts and designated Eh. In a general reversible oxidation-reduction system, the reaction can proceed in either direction according to the conditions. Such a reversible system may be represented as reduced form = oxidized form + electron or

$$[Red] \rightleftharpoons [Ox] + e$$

The reduced form is ionized in accordance with the formula

$$[HR] \rightleftharpoons [H^+] + [R^-]$$

to which the usual mass action equation is applied:

$$\frac{[H^+][R^-]}{[HR]} = k_d$$

The total concentration of reduced form [Red], is equal to the sum of the concentrations of the ionized [R⁻], and unionized [HR], portions,

$$[Red] = [R^-] + [HR]$$

Combining the equations above, the oxidation-reduction systems are shown as

$$[R^-] = [Red] \frac{k_d}{[H^+] + k_d}$$

Then the general electrode equation is applied to the equilibrium:

$$Eh = E_0 + \frac{RT}{F} \ln \frac{[Ox]}{[Red]} - \frac{RT}{F} \ln \frac{k_d}{[H^+] + k_d}$$

where E_o is a constant for the system, R is a gas constant, T is the absolute temperature, and 1n is the natural logarithm. It must be emphasized that Eh is a measure of intensity level and not of capacity.

In nature, the lower limit of redox potential would be present in an environment rich in H_2. On the other hand, the upper limit of redox potential would be found in an environment rich in O_2 without O_2-utilizing systems. Actually, oxygen is the most abundant element in the crust of the Earth with 46.5% by weight, although more than 80% of it is biochemically inert, being bound or buried in silicates, aluminates, and various mineral oxides. Hence only about 0.0001% of the total oxygen in the crust of the Earth, including the hydrosphere, lithosphere, and atmosphere, is free or dissolved O_2. Oxygen present in biomass is 0.0000002% of the oxygen in the crust of the Earth. Oxygen comprises 13% in H_2O, 0.0006% in organic compounds, and approximately 5% in phosphates, sulfates, and carbon dioxide. The redox potentials in most natural ecosystems in contact with the atmosphere range from Eh -50 to $+750$ mV at pH 2.0 to 11.0; in environments isolated from the atmosphere, such as stagnant waters and bottom sediments, they range downward to Eh -400 mV (Figure 10). Organisms that grow at Eh values lower than -350 mV or higher than $+650$ mV are exceedingly rare.

Living organisms differ from one another in their responses to the presence of atmospheric oxygen, and hence they conventionally have been classified according to aerobiosis. Obligate aerobes cannot grow without molecular oxygen. On the other hand, obligate anaerobes are killed by exposure to air because their enzymes must remain in a reduced state in order to function. Such enzymes become oxidized under aerobic conditions, and one or more essential metabolic reactions are prevented when the enzymes are denatured by the oxidation. Facultative anaerobes are those organisms that can grow either in the presence or absence of air; they either have an anaerobic type of metabolism which is not sensitive to oxygen or their metabolism shifts from aerobic to anaerobic depending upon environmental conditions. From the point of energetics in biological systems, however, Brock[53] recently has suggested the following classification. Obligate aerobes require O_2 usually because they are unable to generate energy by fermentation. Anaerobes are those organisms unable to use O_2 as a terminal electron acceptor because they lack cytochromes that transfer electrons to O_2; this kind of obligate anaerobiosis occurs in only two groups of microorganisms: bacteria and protozoans. Facultative aerobes are able to obtain energy either by electron transport phosphorylation or by fermentation. Microaerophilic organisms require O_2 but at pressures lower than 0.2 atm. In this new classification of Brock's, facultative organisms which used to be defined as facultative anaerobes now are defined as facultative aerobes. This nomenclature may be logically based on the development of cytochrome systems in biochemical evolution; it has been shown that cytochromes and other components of the respiratory systems of facultative organisms may be missing or greatly reduced in anaerobically grown facultative bacteria, but synthesis of these components rapidly resumes when oxygen is introduced.[61]

In the hydrosphere, oxygen has the following distribution: H_2O, 99.7875%; sulfate, 0.2103%; carbon dioxide, 0.0011%; $Si(OH)_4$, $Si(OH)_3O$ etc., 0.0006%; borate, 0.0002%; nitrate, nitrite, and phosphate, 0.00011%; dissolved oxygen, 0.00009%; organic compounds, 0.00003%; and all other forms, less than 0.0001%.

The content of dissolved oxygen in natural waters is influenced primarily by the partial pressure of oxygen in the atmosphere. The exchange of oxygen between water and the atmosphere is influenced by many static and dynamic conditions. Hence the change of dissolved oxygen in waters is directly proportional to the difference between the saturation concentration, O_s, and the actual concentration in the water, O, expressed as:

$$\frac{\partial O}{\partial t} = K (O_s - O)$$

where K is the overall absorption coefficient.[62] The value of K depends on the volume of water, V, and the area of the exposed surface, A. The ratio V/A is sometimes referred to as the aeration depth. The rate of aeration may vary under different conditions, being characterized with a transfer coefficient, f. The transfer coefficient also is called the exit coefficient, and is defined by the relation:

$$K = \frac{f\,A}{V}$$

Approximate values estimated for this transfer coefficient, f, in cm/hr are 0.2 to 0.5 for stagnant water, 1.0 to 5.0 for slowly stirred water without appreciable disturbance of the surface, 1.0 to 10 for slowly stirred water with slight surface disturbance, 10 to 50 for rapidly stirred water with regular vortices at the surface, 50 to 500 for rapidly stirred water with irregular vortices and severely disturbed surface, 10 to 37 for regular progressive waves 2.8 to 10.2 cm high in shallow water, 11 to 25 for choppy waves 6.4 to 13.3 cm high in shallow water, and 30 to 200 for shallow and turbulent brooks. Once oxygen is dissolved in water, it is transported to deeper parts of the hydrosphere by various processes.

A simple advection-diffusion model was applied to the deep water of the North Pacific Ocean for the estimations of the oxygen diffusion and the decomposition rate of organic matter.[63] In this model, vertical components of the mixing parameters, eddy diffusivity, and advection velocity, are assumed to be constant, and the horizontal components of the parameters are neglected. Another basic assumption is that the vertical profile of dissolved oxygen is in a steady state. As the deep waters in the oceans may be assumed to be mixed by turbulence and to rise slowly because of their formation almost exclusively in polar regions, the effects of horizontal diffusion and advection on the oxygen concentration in these deep waters can be neglected. The vertical eddy diffusivity (D_z) may not vary appreciably with depth (z). The decomposition of dissolved organic matter in the deep water is very low, and the particulate organic matter may be decomposed appreciably while falling through the water column. If the rate of oxygen consumption through decomposition is proportional to the flux of particulate organic matter, the expression for the rate of oxygen consumption can be written as

$$O = c_1 - c_2 \exp(-W/D_z \times z) + \frac{R_1}{D_z\alpha\,(\alpha - W/D_z)} \exp(-\alpha z)$$

where O is the oxygen concentration, c_1 and c_2 are constants determined from the oxygen concentrations at the lower and upper boundaries, W is upwelling velocity, D_z is the vertical eddy diffusivity, z is the depth, and α and R_1 are constants determined from the vertical distribution of dissolved oxygen. The vertical eddy diffusivity (D_z) has been determined to be 1.3 cm²/sec in the eastern North Pacific Ocean.[63] The mean upwelling velocity is then calculated to be 4 m/year. The rate of oxygen consumption also is calculated to be relatively high in the upper deep layer (1 to 2 km depth) in the northern part of the North Pacific Ocean, while the rate does not vary appreciably with the depth (Table 1). The rate of oxygen consumption at 1 km depth is approximately 49 mg O_2/m³/year in high latitudes, 2.8 to 5.6 mg O_2/m³/year at 1 km depth in tropical regions, and 1.4 to 2.8 mg O_2/m³/year at 3 km depth in all the North Pacific Ocean.[63] As the rate of oxygen consumption in the deep water of the North Pacific Ocean is almost comparable with that in the bottom water of the Antarctic Ocean, the rate of oxygen consumption may be almost constant at 3 km depth in any part of the oceans.

The general distribution of dissolved oxygen is in good agreement with the deep water circulation. The generally high and uniform oxygen content around the Antarctic continent is consistent with the uniform character of the circumpolar water. The deep water in the South Atlantic Ocean rapidly becomes mixed with Antarctic intermediate water and Antarctic bottom water. A small exchange of deep water takes place between the South and North Pacific Ocean. A slow clockwise circulation of the deep water is known in the North Pacific Ocean, and the oxygen content of the deep water decreases by biochemical processes during the long time taken to move from the western to the eastern side of the ocean. The conditions in the Indian Ocean are more or less similar to those in the South Pacific Ocean.

In contrast to the marine oligotrophic environment, dissolved oxygen in eutrophic lakes, or hypereutrophic lakes as their extreme, may be highly dynamic due to bursts in the biological activities of photosynthesis and respiration. As lake communities are composed mostly of obligate aerobes, microorganisms which actively modify aerobic conditions in such aquatic environments of oxygen instability threaten even their own existence. When phytoplankters die after an algal bloom, they are decomposed by heterotrophic microorganisms. When a large amount of algal material starts to be decomposed, the microorganisms use up much of the dissolved oxygen. As the oxygen concentration in the water decreases below the level required by fish, these animals cannot continue breathing and soon die. Such critically low concentrations of dissolved oxygen and unfavorable temperatures cause the "summer kill" of fish (Figure 11). This contrasts with another catastrophic destruction of aquatic life; the "winter kill". When ice covers shallow waters during winter, large amounts of organic matter in the bottom of eutrophic waters may exhaust the dissolved oxygen of the unfrozen water resulting in the mortality of fish and other organisms.

These unfavorable oxygen conditions, microaerobic or anaerobic, are modified partly by oxygen transportation through physicochemical processes in lake water. The dynamics of dissolved oxygen in a freshwater environment were analyzed during an algal bloom in Lake Kasumigaura, Japan, during a 24-hr windless condition.[64] Phytoplankters comprising the algal bloom were composed predominantly of *Microcystis aeruginosa*, *Anabaena spiroides*, *Cryptomonas erosa*, *Coscinodiscus rothii*, *C. lacustris*, *Carteria cordiformis*, *Scenedesmus quadricauda*, *S. actiformis*, *Pandorina morum*, *Eudorina unicocca* and *Coelastrum microporium*. Major groups of the biological agents (as indicated by ATP) responsible for the production and consumption of dissolved oxygen were phytoplankton (as indicated by chlorophyll a) and bacteria, respectively. The statistical analyses showed that these organisms exhibited no significant vertical stratification, with averages and standard deviations as follows:

ATP $= 6.2 \pm 3.2 \, \mu g/\ell$
Chlorophyll a $= 150 \pm 41 \, \mu g/\ell$
Bacteria $= 1.2 \times 10^7 \pm 0.8 \times 10^7/m\ell$

In situ dissolved oxygen started to increase in the shallow water (above 60 cm) immediately after dawn, whereas it was decreasing in the deeper waters (below 60 cm) in the early morning (Figure 12). Dissolved oxygen reached the maximum at noon, when the concentration at each depth throughout the water column (except for the bottom) was almost doubled compared to the concentration at dawn. The maximum concentration of dissolved oxygen in the upper water was 190% of saturation, when pH and Eh were 8.90 and 294 mV, respectively, at 32.0°C. Thereafter, dissolved oxygen decreased almost continuously until the next dawn. Absorption of atmospheric oxygen or liberation of dissolved oxygen through the water surface is directly proportional to the difference between saturation concentration and *in situ* concentration.[62] The daily oxygen budget at the lake surface was determined to be 2.1 mg O_2/cm^2 liberated per day. The oxygen consumed by microorganisms (mg O_2/ℓ/day) was 6.2, 4.7, 3.3, 3.2, and 3.9 at each zone of 0 to 20 cm, 20 to 40 cm, 40 to 60 cm, 60 to 80 cm, and 80 to 100 cm, respectively. The eddy diffusion coefficient (cm^2/sec), D_z, was calculated as 6.7, 17.4, 14.9, 1.1, and 2.3 at each zone of 0 to 20 cm, 20 to 40 cm, 40 to 60 cm, 60 to 80 cm, and 80 to 100 cm, respectively. These D_z values are similar to those commonly measured directly in the surface layers of the coastal region of Japan. The daily oxygen transportation (mg O_2/ℓ/day) to the deeper layer was 10, 7.4, 4.9, 0.52, and 0.48 at each zone of 0 to 20 cm, 20 to 40 cm, 40 to 60 cm, 60 to 80 cm, and 80 to 100 cm, respectively. Thus, during the windless condition, major parts of the dissolved oxygen produced during the algal bloom were liberated from the lake surface or were respired by microorganisms, as low eddy diffusion coefficient values at deeper layers within the euphotic zone prevent the efficient supply of dissolved oxygen to the dysphotic zone. This mechanism seems to minimize the direct influence of highly unstable oxygen conditions in the euphotic zone on the dysphotic zone.

The same kind of analyses on the dynamics of dissolved oxygen was made during typical wind conditions over the four seasons (Figure 13). The daily oxygen budget at the water surface showed that oxygen dissolving into lake water from the atmosphere was 0.66, 1.18, and 0.65 mg O_2/cm^2/day in November, January, and April, respectively. On the other hand, oxygen liberated from the lake water was 1.16 mg O_2/cm^2/day during the blue-green algal bloom in August. The diffusion coefficient for the vertical transportation of dissolved oxygen by eddy diffusion during a typical local wind in any season was only one order of magnitude greater than that for the windless condition. Thus, the windy condition still favors stability of oxygen conditions in the dysphotic zone of Lake Kasumigaura.

The same type of profile of dissolved oxygen can be observed in eutrophic water masses of the marine environment. For example, the water masses in a gyre of coastal water in the inner part of Tokyo Bay were vertically stratified throughout the summer and early autumn (Figure 14). Three water layers — surface, intermediate, and bottom were distinguished. The surface layer was less saline as a result of freshwater inflow. The density of phytoplankton biomass was high, and a compensation depth was always found near the bottom of this surface layer. The occurrence of active photosynthesis was reflected by oversaturation of dissolved oxygen (up to 180%) and high pH (up to 8.8). In this red tide, a large number of copepods, predominantly *Acartia clausi* and *Oithona nana*, were found grazing on phytoplankters, primarily *Eutreptiella, Gymnodinium, Gyrodinium, Exuviella,* and *Prorocentrum.* The biomass of plankton in the intermediate layer was less than 10% of that in the surface layer, and dissolved oxygen was undersaturated, usually between 30 and 60%. Phytodetritus and fecal pellets of zooplankton often formed large aggregates, measuring 7 mm below the boundary between the surface and intermediate layers. The bottom layer was characterized by concentrations of dissolved oxygen of less than 30% saturation. Distribution of oxygen was not homogeneous, and water patches containing hydrogen sulfide were commonly found in this layer.

The basic processes whereby dissolved oxygen is consumed in natural waters are classified into seven categories:[156]

1. Microbial oxidation of organic matter
2. Bacterial respiration
3. Respiration of higher organisms including phytoplankton and zooplankton
4. Microbial oxidation of hydrogen gas or methane gas evolved in anaerobic bottom deposits
5. Microbial oxidation of hydrogen sulfide, ferrous compounds, ammonium, nitrite, thiosulfate, etc.

6.   Abiotic oxidation of inorganic compounds
7.   Abiotic oxidation of organic compounds

Thus living organisms reduce the redox potential of their environment not only by the consumption of dissolved oxygen but also by the production of metabolic products. Hydrogen sulfide is a common product of anaerobic metabolism and reduces the redox potential down to -300 mV. Natural environments of low redox potential include muds and sediments of lakes, rivers, and oceans, and bogs and marshes where micro-aerobic and anaerobic microorganisms are predominant.

The rate of abiotic oxidation is highly dependent on environmental oxygen concentration and is enhanced by disturbance of the surface sediment and exposure of the anoxic layer to oxygen (Figure 15). On the other hand, the respiration rate of a mixed population of aerobic benthos has been shown to decrease more slowly with reduction of available dissolved oxygen, until the oxygen concentration falls below a critical concentration of approximately 1.4 mg O_2/ℓ.[67] Below this critical concentration, the response in microbial respiration to the availability of dissolved oxygen changes greatly, possibly due to the interchange of microbial flora from aerobic to microaerophilic. Mortimer[68] also observed that the rate of oxygen depletion in a stratified lake was rather constant until the oxygen concentration of the hypolimnion decreased to 1.9 mg O_2/ℓ, at which point the rate of oxygen consumption started to decrease rapidly.

This critical O_2 concentration, below which oxygen consumption decreases rapidly, was estimated for a pure strain of marine bacterium to be as low as 0.43 mg O_2/ℓ, and the average consumption rate was 2.1×10^{-11} mg O_2/cell/hr.[69] This rate also was determined to be independent of the ambient concentration of dissolved oxygen within the range of 0.43 to 17.84 mg O_2/ℓ. These differences in the critical O_2 concentrations may be due to differences in the threshold concentrations of oxygen consumption by a pure strain of marine bacterium or by mixed populations of microbenthos in both freshwater and marine environments. The minimum oxygen concentration required for aerobic energy-generating processes is determined biochemically to be approximately 0.2%, which corresponds to 1% of the oxygen concentration in the present atmosphere (20%). This threshold of oxygen concentration (0.2%) is almost equivalent to that established for the mixed population of microbenthos, which possibly transport oxygen into their cells by simple diffusion. On the other hand, the oxygen threshold concentration of the marine bacterium is less than one third that of the microbenthos. This low threshold concentration in the oxygen uptake system of a procaryote may be of great advantage over other types of aquatic organisms in microaerobic environments.

The oxygen consumption rate of a pure strain of a marine heterotrophic bacterium[69] is approximately the same as those of marine populations of free-living bacterioplankters in different water masses (Figure 16). The water masses of the surface layer of Tokyo Bay (35° 32' N, 139° 54' E), Aburatsubo Inlet (35° 09.2' N, 139° 37' E) the subarctic Pacific water (44° 00' N, 154° 00' E) and the western central North Pacific water (28° 25' N, 145° 00' E) are typically eutrophic, mesotrophic, oligotrophic, and ultraoligotrophic, respectively.[70] Microbial oxygen consumption in these four water masses was compared during the summer, when environmental conditions are optimal. If the respiration values had been compared at different seasons, oxygen consumption could be affected by many factors such as changes of thermal kinetics in biological reactions and interchange of microbial groups having different biological processes. Measurements of microbial respiration were done on seawater samples prefiltered with sterilized Whatman® glass fiber filters GF/C (diameter 45 mm; pore size 1 μ) to eliminate zooplankton, phytoplankton and microbial aggregates. Thus cell numbers were determined only for free-living bacteria in the seawater. Respiration by bacterioplankton (in mg O_2 per bacterial cell per hr) was determined to be $(4.7 \pm 2.4) \times 10^{-12}$ in the ultraoligotrophic or oligotrophic water masses of the oceanic regions, $(1.2 \pm 0.7) \times 10^{-11}$ in the mesotrophic coastal water mass of the inlet, and $(4.8 \pm 3.6) \times 10^{-12}$ in the eutrophic water of the bay. Oxygen consumption by bacterioplankters may be regulated to a relatively constant rate, which is only slightly affected by the degree of eutrophication.

Oxygen is the most common electron acceptor in the electron transport systems of aerobic organisms. Some organisms can use other electron acceptors in the process of anaerobic respiration. Sulfate is known to be an important reservoir of biochemical oxygen in nature. Sulfate-reducing bacteria use the oxygen in sulfate to oxidize organic compounds or molecular hydrogen by the following reactions:

$$H_2SO_4 + 2(CH_2O) \rightarrow 2CO_2 + 2H_2O + H_2S$$

$$H_2SO_4 + 4H_2 \rightarrow 4H_2O + H_2S$$

Sulfate-reducing bacteria are distributed in anaerobic waters and bottom deposits of the hydrosphere. Hydrogen sulfide produced by these bacteria causes oxygen depletion in stagnant waters of both freshwater and marine environments. Hydrogen sulfide-oxidizing bacteria, on the other hand, are responsible for the formation of sulfate. Nitrate is another important reservoir of biochemical oxygen in nature. The low concentration of nitrate in natural waters and the accumulation of ammonium are reflections of the ease with which nitrate is reduced by many aquatic microorganisms. The ferric ion (Fe^{3+}) is used by certain bacteria as an electron acceptor, being reduced to the ferrous ion (Fe^{2+}) in the hydrosphere. Some other bacteria can use carbon dioxide and certain organic compounds as electron acceptors; the methanogenic bacteria are a well-known group of strict anaerobes that use carbon dioxide as an electron acceptor and reduce it to methane.

It has long been known that many aquatic organisms can survive for a certain period in the absence of free oxygen. Certain representatives of protozoans, nematodes, earthworms, leeches, immature stages of insects, mollusks, fishes, and others exhibit this ability. During the absence of free oxygen, a certain amount of energy may be released by the splitting of carbohydrates into reduced substances, e.g., in some mollusks and certain oligochaetes glycogen is used in excess when they are subject to anaerobic conditions.

V. SURFACE TENSION

The surface tension of a liquid is an index of the force field at its exposed surface. It is usually expressed in dyn/cm. Its dimension corresponds to g/sec^2 or to the unit of free surface energy in erg/cm^2. This surface tension occurs because surface molecules are attracted downward only; upward attraction is lacking because there are no water molecules above the surface molecules. Accordingly, a surface tension is produced which acts inwardly, i.e., the molecules act as if they formed a tightly stretched membrane over the water. This mechanism serves as a mechanical support for certain objects just under the water surface of the hydrosphere. Molecules in the interior of a water mass do not exhibit this phenomenon, because there they are attracted to each other in all directions and are thereby neutralized. Surface tension diminishes with a rise in temperature (t) and increases with salt content (Cl) as follows: surface tension (dyn/cm) = $75.64 - 0.144t + 0.0399Cl$. Thus the surface tension of most natural waters is slightly greater than that of pure water at the same temperature. The surface tension of seawater is 73 dyn/cm at 20°C.

On the other hand, surface tension is lowered by organic compounds in solution. These organic compounds are called "surface active substances" or "surface tension depressants". These substances do not form homogeneous solutions in water, but occur in very much higher concentration at the surface. The surface tension depression, then, is related strictly to the ratio of the concentration of these substances in the surface to that in the whole water column. All these surface-active substances are known to have a common molecular pattern, with one part of the molecule being hydrophobic and the other hydrophilic. The hydrophobic moiety consists of a hydrocarbon chain. The hydrophilic part may have a negatively or positively charged ion such as a carboxyl group or an ammonium group, or it may be a neutral group such as an ether link.

Most microorganisms grow best at surface tensions ranging between 45 and 65 dyn/cm. A reduced surface tension tends to increase the concentration of dissolved nutrients at the cell-water interface and also the collision factor between organisms and substances in solution or suspension. Thus a reduced surface tension facilitates the excretion of waste products from cells into their environment. However, a surface tension below 40 dyn/cm tends to upset the osmotic equilibrium by adversely affecting transport systems through damaging cell walls and membranes and promoting leakage of essential metabolites.

Uncolored freshwaters, which are poor in plankton, exhibit no depression of surface tension below that of pure water with a surface tension of 72.7 dyn/cm at 20°C. However, surface active materials are produced by the vegetation of lakes and rivers. These materials may lower the surface tension as in colored bog lakes, where surface tensions range between 53 and 73 dyn/cm and average 67 dyn/cm. These surface tensions are in the desirable range for the activities of most freshwater organisms.

The greatest effects of surface tension are believed to be on surface films, the uppermost layer of water, and the neuston therein.[71] Surface tension also greatly affects the ability of living microorganisms to attach to solid surfaces. When certain conditions are satisifed, the surface film on natural waters serves as a mechanical support for miscellaneous particulate materials including living organisms. These organisms are called neuston. The term "neuston" was proposed originally to include bacteria, euglenids, chlamydomonads, amoebas, and other minute plants and animals inhabiting the surface films of small ponds and puddles.[72] The definition has been extended to include all organisms associated with the surface film. The neuston found in the upper surface of the film comprise the epineuston and those related to the lower surface comprise the hyponeuston. Both of these groups consist of a great variety of plants and animals.

The lower part of the surface film supports representatives of the following groups of hyponeuston: bacteria, phytoplankters, protozoans, hydras, planarians, ostracods, cladocerans, insect larvae, insect pupae, and pond snails. Most of these organisms spend part or all of their life cycle suspended from or creeping on the lower side of the surface film or swimming close to the water surface; they feed on organic detritus or on each other. Hydras, planarians, and pond snails are not true representatives of the hyponeuston but are regarded merely as occasional visitors. On the other hand, the upper surface film supports representatives of the following groups of epineuston: bacteria, phytoplankters, protozoans, water striders, broad-shouldered water striders, water measurers, hebrids (Hebridae), mesoveliids, whirligig beetles, springtails, and certain spiders. The characteristic organisms of the hyponeuston and epineuston can thrive only in small, stagnant, or sluggish waters, where they are suspended easily in the surface film. This leads to the opinion that this neustonic complex is possible only in calm ponds and puddles and cannot thrive in the open parts of lakes, seas, or oceans. Therefore, the neuston community in the marine environment commonly is subdivided, based on a more approximate depth distribution, into the euneuston, facultative neuston, and pseudoneuston.[73,74] Euneuston are those organisms with maximum abundance in the surface during both day and night. Facultative neuston are organisms which concentrate at the surface only during certain times, commonly during darkness. Pseudoneuston are those organisms abundant in the surface but with maximum densities in deeper layers.

The abundance of life in the topmost layer of the marine environment has become evident. In comparison with the underlying waters, this surface layer contains greater biomass of each of the following groups: floating macrophytes, phytoplankters, metazoans, protozoans and bacteria.

The density of bacterioneuston has been shown to average two or three orders of magnitude greater than that of bacterioplankters.[75] Bacterioneuston in different parts of the oceans are characterized by a great species diversity, a prevalence of colored forms, and a definite selectivity of genera (Figure 17). The presence of these bacterioneuston communities may be especially favored by the accumulation of organic materials created by the peculiar properties of the air-water interface. The group of heterotrophic bacteria, which are predominant in the euphotic zone are also predominant in the surface film; foam formation is considered to be a major mechanism linking the bacterioplankton and bacterioneuston.[77]

Bacterioneuston depend on organic materials accumulated in the surface film. Abiotic organic materials influx into the surface film from various sources. Major fractions originate from aquatic organisms as products of their metabolism and decomposition. Other organic materials enter from river runoff and precipitation from the atmosphere. These materials are in particulate, colloidal, and dissolved forms. Land insects carried by the wind into aquatic environments cannot be neglected as sources of large particles of detritus, especially in coastal regions. Dead insects contain large amounts of organic compounds which serve as sources of energy and material for fishes and invertebrates inhabiting the surface layer. Pollen, spores, and other organic particles also fall steadily on the water surface from the atmosphere. Especially in pelagic regions, abiotic organic materials in the water surface originate from the remains and excretions of aquatic plants and animals. The antirain phenomenon must be important in the surface accumulation of such organic debris from aquatic organisms; this phenomenon occurs when decaying bodies of some crustaceans and other plankters become buoyant and rise to the water surface. This contrasts with other types of dead plankters which become negatively buoyant and fall like rain through the water column to the bottom. Antirain leads to the concentration of fragments and whole corpses of dead aquatic organisms in the surface film and in foam. The concentration of dead plants and animals becomes greater in the surface film than below, and these organisms decompose further, supplying neustonic detritivores with fresh detritus of high nutritional value. Some fractions of this fresh detritus are grazed directly by animals inhabiting the water surface and some are assimilated by neustonic microorganisms. Gas bubblings are known to be another major mechanism for the accumulation of organic materials by the processes of adsorption, aggregation, and redistribution of abiotic organic materials which the bubbles carry to the water surface.[78,79] By this mechanism, foams are constantly formed with abiotic organic materials rising from deeper layers to the surface. Abiotic organic materials in the water column also are adsorbed on the surfaces of detrital particles of dead bodies or aggregates in the surface film. Formation of these particulate organic materials stimulates the activities and multiplication of heterotrophic microorganisms by increasing the availability of nutrient-rich microenvironments.

Excretion of organic materials from phytoneuston also may contribute greatly to the accumulation of organic debris in the surface film. Since live diatoms, dinoflagellates, and blue-green algae do not usually float up and aggregate just below the surface film, except in the case of red tides in coastal regions, the role of macrophytes is more important in this respect. Benthic sargassum species float to the surface when detached from the sea floor. Floating algae of this family occur almost exclusively throughout tropical, subtropical and temperate waters (Figure 18).[80,81] The low mineral content of the thallome and the well-developed air bladders are responsible for the high buoyancy of hyponeustonic species of *Sargassum.* These plants are well-known as the dominant form of floating algae in the Sargasso Sea. Quantitative estimations of hyponeustonic algae have been made in the Sargasso Sea, where the standing stock of sargassum algae is between 1.8 and 2.35 g/m^2 [82] or between 0.9 and 2.5 g/m^2.[80] The distribution of these plants is influenced greatly by wind currents. During strong winds, patches of these plants accumulate in convergence zones. The standing stock of these hyponeustonic plants is of the same order of magnitude as that of phytoplankton in the Sargasso Sea. Incidentally, the standing crop of phytoplankton carbon beneath a unit area of the euphotic zone in the marine environment is in the range of 1 to 2 g C/m^2 for most oceanic regions and 2 to 10 g C/m^2 for coastal or other fertile seas for most of the year, except during winter when not more than 0.5 to 1 g C/m^2 would be expected.[83] As 1 g of phytoplankton biomass corresponds approximately to 0.4 g of phytoplankton carbon, the contribution of floating algae in the sea to the dynamics of organic materials is very appreciable.

Among neustonic protozoans, tintinnids are regarded as important components of the marine hyponeuston. Radiolarians and foraminiferans also are regarded as widespread constituents of the oceanic hyponeuston.[73]

Microplankton form the next link of the food chain because they graze exclusively on protozoans and bacteria. Tsyban[75] has shown a linear relationship with high correlation between the bacterioneuston and small invertebrates inhabiting the surface layer. The commonest forms of these small hyponeustonic invertebrates are rotifers, cladocerans, copepods, and larvae of polychaetes, gastropods, lamellibranchs, cirripedes, and echinoderms. The absolute or relative abundance of these species among hyponeustonic communities varies considerably with environmental conditions.[73] Fish larvae and fish fry may form the next link of the food chain in neustonic communities.

The distribution of neuston generally is affected by surface currents and winds which sweep epineustonic forms along the top of the water. Waters sinking in convergence zones are characterized by a greater density of aquatic animals and other various floating substances, both autochthonous and allochthonous. Those organisms having high buoyancy resist the sinking currents and remain in the water surface of convergence zones, where they adapted morphologically, physiologically, and behaviorally as neuston. Such convergence zones, although unfavorable for the development of grazing food chains, are the best environments for development of neustonic detritus food chains composed of all trophic levels from bacteria to fishes (Figure 19). In contrast, grazing food chains are prevalent in divergence zones where a rich phytoplankton community develops because of upwelling of deep waters rich in nutrients. In these areas, the phytoplankton populations are grazed by herbivorous zooplankton, and the herbivores are eaten by carnivores.

There is a claim that while the population densities of neustonic organisms may be high, the total biomass of neuston in the water column is small because only a very thin layer is occupied by this community. The detritus food chains in the surface film, however, are comprised of bacteria, protozoans and small invertebrates, all developing in a nutrient-rich medium of organic debris. Characteristic of these organisms is their rapid rate of reproduction; the production rate of their biomass may be very high although their standing stock may be small in the total water column. Accordingly, these minute neustonic organisms can be skimmed as a productive food source by organisms in the underlying water.

VI. HYDROSTATIC PRESSURE

The biosphere does not end where light ceases, because there is a flow of energy downward through the falling of abiotic particles from the euphotic layer to the depths and through the active vertical migrations of zooplankton and nekton. Organisms living at great depths are thus exposed to high hydrostatic pressures.

As the depth-pressure gradient in the hydrosphere ranges from 0.099 to 0.105, depending upon salinity, temperature, latitude, compression, and other factors affecting seawater density, the hydrostatic pressure in the Vitiaz Deep is approximately 1160 atm at 11,034 m depth, the greatest known depth of the oceans. In the freshwater environment, only two lakes, Lake Baikal and Lake Tanganyika, are known to have maximum depths over 1000 m.

The fundamental relation of pressure and temperature has been expressed in the ideal gas law as $PV = nRT$, where P is the pressure, V is the volume, n and R are constants, and T is the absolute temperature. The law can be applied qualitatively to biological systems, but not quantitatively. The theoretical basis whereby pressure affects rates of ordinary chemical reactions is given in the review by Stearn and Eyring.[84] The specific rate constant for any elementary chemical process can always be written as:

$$K_p = K_o e^{-p\Delta V/RT}$$

where K_p is the rate under pressure p, K_o is the rate at zero pressure, and ΔV is the volume of the activated complex minus the volume of the reactants. In this equation, when p is taken in atmospheres and the gas constant R in cm^3 atm (82.07 mℓ/mol), ΔV is expressed as cm^3/mol at absolute temperature T. The prediction of the pressure effect on the rates requires only the estimation of ΔV. Moreover, in any biological system the pressure can modify the viscosity of protoplasm and alter the solution of macromolecules. Such effects may be directly responsible for changes in the rates of metabolic processes altered by pressure.

A series of classical studies of these relationships have been made on the mechanism of light emission by luminous bacteria. Luminescence intensity of a psychrophilic bacterium, *Photobacterium phosphoreum*, changes as a function of pressure at different temperatures (Figure 20). In the luminescent system, molecular oxygen, luciferase, and reduced luciferin interact to emit a quantum of light energy. The activation of the luciferase is temperature dependent and has a ΔE of 17,000 cal. A temperature rise results in an increase in the light intensity emitted up to some maximal value, with the sharp decrease beyond this maximal temperature resulting from a protein denaturation reaction; i.e., above this critical temperature a different master reaction is controlling the series of reactions from that which is controlling below the critical temperature. However, when pressure is applied to this luminescent system at temperatures above the optimum, the effect of temperature on the denaturation of luciferase is counteracted by the effect of hydrostatic pressure and an increase in light intensity takes place. Accordingly, the master reaction involves a volume increase associated with an inactive state of the enzyme in the range above the optimal or critical temperature, whereas the master reaction may involve a volume increase associated with an active enzyme state below the critical temperature. Thus a reciprocal interaction of hydrostatic pressure and temperature on biochemical reactions, metabolism, and growth could be expressed by LeChatelier's theorem; i.e., when an external force is applied to a system in equilibrium the system will shift in such a way as to minimize the effect of the applied force. The same kinds of reciprocal effects of pressure and temperature are expected in any biological system involving reversible changes in volume. As an extreme example some sulfate-reducing bacteria, possibly *Desulfovibrio*, were isolated from oil and sulfur wells at depths exceeding 4000 m, where the *in situ* temperature was 60 to 105°C and the hydrostatic pressure was 400 atm.[86] One culture was grown experimentally under conditions of 104°C and 1000 atm, with sulfate in the medium being reduced to sulfide.

Most organisms inhabiting the shallow water environment or the terrestrial environment grow best at atmospheric pressure. Their growth is inhibited by increased hydrostatic pressure. Organisms which grow poorly or not at all at pressures higher than 300 to 400 atm are termed barophobic, meaning they dislike high pressure. Lethal effects caused by high pressure are believed due to inactivation of cellular proteins.

On the other hand, marine organisms show a somewhat greater resistance to pressure on the whole. There may be a correlation between the depth which they inhabit and their ability to grow under pressure.[87] Marine bacteria (*Achromobacter fischeri*, *A. harveyi*, *A. thalassius*, *Photobacterium splendidum*, *Bacillus cirroflagellosus*, *Micrococcus infirmus*, *Pseudomonas pleomorpha*, and *Vibrio hyphalus*) obtained at or near the sea surface tend to resemble terrestrial organisms in their response to pressure, whereas some species such as *Bacillus submarinus* and *B. thalassokites* isolated from 5840 m in the basin north of Bermuda (30° 28′ N, 65° 22′ W), where the pressure approximates 600 atm, grew well under 600 atm at 30°C and 40°C and were obviously more resistant to the effects of 400 and 500 atm than terrestrial bacteria (Table 2). Mixed microflora, from mud freshly collected from the ocean floor at depths up to 3600 m, multiplied rapidly under pressures between 300 and 600 atm at 20, 30, and 40°C. Growth of some bacteria was greater under 400 to 600 atm than at normal atmospheric pressure. Certain sulfate-reducing bacteria, isolated from oil-well brines a few thousand meters below the surface of the Earth, are also metabolically more active when compressed at 400 to 600 atm than at 1 atm. Therefore, ZoBell and Johnson[87] coined the term "barophilic" to characterize such organisms which possess the ability to grow and carry on metabolism as well or better under increased pressure.

At present, living organisms have been detected from all depths of the hydrosphere. In 1950 during the Galathea Expedition, the occurrence of living bacteria was demonstrated for the first time on the ocean floor of the Mariana Trench, which includes the deepest parts of the ocean (Table 3). Piccard[48] also confirmed the existence of a native fish living on the floor of the Mariana Trench during a bathyscaphe dive to 10,960 m. Piccard described it as an endemic flat fish and not a fish which had accompanied the bathyscaphe during the descent.

Eurybaric organisms are able to grow over a wide range of pressures from 1 atm to 500 atm or higher. Barotolerant organisms survive for prolonged periods at high pressures from 500 up to more than 2000 atm, but they may not be able to grow at such high pressures.[89]

As expressed by the theory on the pressure effect, these barotolerant and eurybaric organisms may exhibit biological activities which are one or two orders of magnitude slower than those in shallow depths.[90-94] Such low activities may be achieved both by autochthonous and allochthonous organisms that are exposed to dilute concentrations of biochemically refractory organic materials in the deep oceans. On the other hand, moderate rates of activity that may be mostly due to barophilic organisms have been detected sporadically at *in situ* conditions of the deep oceans.[24,54,95-99] These barophilic bacteria, which may occur only very sparsely in the deep sea, can multiply with generation times of about 10 hr when supplied with abundant nutrients; e.g., the generation time of obligate barophiles was 8 hr when measured at 5200 m depth.[95,97] The nutrient enrichment of the water sample at 5200 m was checked *in situ* by a deep-sea camera.[97] The predominant bacteria which multiplied with a generation time of 8 hr were obligate barophiles because they could not grow at 1 atm.[95] However, eurybaric bacteria may have generation times of 1 to 10 days.[43] The depression of propagation of barophilic bacteria in deep-sea conditions, which makes them only a minor group in the resident microbial flora, may be connected with the refractory nature of organic materials in the deep oceans inhibiting usual microbial growth. The lack of energy sources for long periods has been shown to permit the evolution of slowly metabolizing bacteria. A marine psychrophilic bacterium, *Vibrio* Ant-300 from the Antarctic Ocean, decreases its cell size during starvation.[100] The cells appeared as small rods and small cocci after 1 and 6 weeks of starvation, respectively (Figure 21). Size reduction is most rapid during the first 2 days of starvation and continues for 3 weeks. After 2 weeks of starvation, 100% of the viable cells are less than 1 μm. After 3 weeks of starvation, 100% of the viable cells are less than 0.6 μm. Nonstarved *Vibrio* cells are typical Gram-negative cells and show no unusual structures (Figure 21a). The small starved cells (Figures 21b and 21c) appear smaller and roughly spherical. Their ultrastructure is generally the same as in nonstarved cells except for an enlarged periplasmic space containing stainable material. No unusual membrane structures are observed. After 5 weeks of starvation, cells are less than 0.4 μm. With the addition of nutrients, these cells start to grow without a significant lag period and with a generation time of 6 hr. Small coccoid cells immediately begin to increase in size, and they regain their rod shape within 48 hr. Thus the rate of deep-sea microbial activities also may be related to the availability of energy in the system.[99] Most autochthonous bacteria associated with sediments and water of the deep sea may have very slow metabolic rates. These bacteria have evolved to be slow metabolizers due to the age of the water mass and the lack of energy sources, whereas some deep-sea microorganisms in constant contact with an energy source may have moderate metabolic rates.

The quantity of plankton generally decreases rapidly with increasing depth. The decrease in standing stock usually is most marked from 100 to 200 m and from 500 to 700 m, although the quantity of plankton in deep layers always is related to the quantity of plankton in the overlying surface layer (Figure 22).

During the daytime, grazing by herbivores is minimal because many of these animals stay below the euphotic layer. Hence the phytoplankton can produce as much organic material as possible during the photosynthetic period with minimal grazing pressure. During the evening, the herbivorous animals migrate towards the surface and feed on the phytoplankton. This mechanism permits the greatest primary production by phytoplankton and the greatest transfer to animal biomass. In the early morning, the animals descend to their deeper day-depth where they remain until the following evening. By staying at deeper locations during the nongrazing period, animals can maintain low metabolic activities due to higher hydrostatic pressure and lower temperature, thereby supporting higher standing stocks by minimizing energy expended in metabolism.

Many animals perform short-term migrations over considerable depth ranges. Some of them perform vertical migrations over distances of 2000 to 4000 m. Others, such as *Calanus cristatus*, feed in the surface layer during certain periods of their life and then descend to depths of more than 500 to 1000 m or even down to 2000 m. Animals descending from the surface layer encounter their predators in deeper layers, where the predators feed on them and their remains. These predators, in turn, descend to still greater depths and become the prey of animals living in the deeper layers. Thus organic materials produced in the surface layer are actively carried to the depths of oceans by these vertical migrations of interzonal and deep-sea species; this process has been termed a "ladder of migrations"[101] or an "ecological ladder" (Figure 23). This active transport of organic materials is believed to be the most important way of supplying food from the photosynthetic layer to the ocean bottom.

These vertical migrations by animals are only possible through their active responses to changes in hydrostatic pressure. It is well-known that the depth regulation of zooplankton performing extensive vertical migrations is influenced by diel changes in light intensities and by activity cycles. This may allow sinking or active movement of animals to considerable depths in response to increasing daylight. On the other hand, many marine animals can regulate their depth through pressure responses. As these depth regulating mechanisms operate independently of light, pressure responses are particularly useful where light penetration is restricted, and they have obvious advantages in restricting migrations within precise depth limits.[102]

The swim-bladders in teleosts and the ampullae of Lorenzini in elasmobranchs are sensitive to local pressures. Statocysts of some invertebrates also may be concerned with perceiving hydrostatic pressure changes. Potentials across the surface of decapod crustaceans vary by approximately 10 mV under the influence of pressure changes of about 3 bar. The whorled lamellate arrangement in the apical region of *Pleurobrachia* and other ctenophores reacts to touch, turbulence, and pressure changes.

Within the limits of their migrations, animals must satisfy their energy requirements. An approximation of the food required (Figure 24) during the growth of a common filter-feeding copepod, *Calanus finmarchicus*, can be compared with *in situ* concentrations of particulate food in the oceans. The amount of food required for the growth of animals ranging between 0.005 to 5.0 mg (wet weight) can be determined from the following equation of Winberg:[103]

$$\% \text{ body weight required/day} = [10^{1/t(\log W_2 - \log W_1)} - 1] \times 100$$

where W_1 and W_2 represent the animal's weight at the beginning and end of the period (t). For *Calanus*, t = 100 days for an animal to develop from a nauplius to a stage V copepod, the change in biomass during this period being 1000-fold. Assuming a food assimilation efficiency of 80% and growth increases of 0.7% and 7% per day at 5° and 10°C, the quantity of carbon required for each weight of animal is given in Figure 24. The grazing of particulate materials by certain zooplankters does not continue below certain critical concentrations of carbon.[104] For a growth rate of 7% per day, the quantity of particulate carbon in the water must be greater than 100 μg/ℓ and, for the smallest copepods at 10°C, the carbon concentration should be more than 250 μg/ℓ. These estimates give the minimum requirements, because the 80% food assimilation assumed in the calculations employed here should be considered as a maximal figure.

Critical carbon concentrations required by some zooplankters for grazing are approximately 70 μg/ℓ for *Calanus*, 58 μg/ℓ for *Pseudocalanus*, 131 μg/ℓ for *Euphausia pacifica*, and 74 to 79 μg/ℓ for a mixture of *Calanus pacificus* and euphausiid furcilia. In all these cases, the carbon concentrations at which grazing starts are within the range found in the euphotic zone. On the other hand, these critical carbon concentrations are above those generally found in deep waters, where particulate carbon concentration is less than 20 μg/ℓ.

On the other hand, these figures do not take into account that some particulate materials may occur as local aggregates. This would tend to improve grazing for filter feeders by providing high concentrations of carbon at discrete locations within a volume of water where the average content of particulate carbon may be very low. It has been suggested convincingly that fecal pellets contribute to local accumulations of metabolizable organic materials in the deep-sea environments. In the Sargasso Sea, which is one of the typical ultraoligotrophic regions, the net vertical flux of materials collected by a sediment trap at 5367 m was 46.0 mg/m²/day.[105] The major biogenic particulate constituents were fecal pellets, planktonic foraminiferal tests, radiolarians, pteropod shells and fragments, copepod molts, *Thoracosphaera* (a coccolithophorid), silicoflagellates, diatoms, free coccoliths, and free pigments. Two kinds of fecal pellets (green and red) were collected which differed in respect to their content of pigments, skeletal carbonate particles, and clay minerals. Green fecal pellets are believed to be transported directly from the surface layer. These green fecal pellets or their fragments may be digested by deep-sea zooplankters since they are a nutritionally attractive food for these detritivores. By this process the organic content will decrease and undigestible clay and skeletal materials will increase at each step. On the other hand, the red fecal pellets show significant differences in skeletal and organic concentrations and different states of preservation of pigments and pigment-like particles. Thus these red fecal pellets, containing less organic material and more clay particles, are speculated to be produced in the deep water by coprophagic zooplankters.

These fecal pellets as well as crustacean molts, animal carcasses, and large phytoplankton cells comprise appreciable fractions of the large particles which contribute almost exclusively to the vertical flux of organic materials in the marine environment; however, these large particles form only minor fractions of the standing stock of total particles in the water column.[106] Sinking rates of these particles have been shown to follow Stokes' law;[106,107] thereby larger particles sink more rapidly and often reach the ocean floor in great depths. The sinking rates of these particles through a nonturbulent water column have been calculated using Stokes' equation:

$$V = \frac{2}{9} \, gr^2 \, \frac{\rho - \rho_0}{n}$$

where V is the velocity in cm/sec, r is the particle radius in cm, g is the acceleration due to gravity (980 cm/sec^2), ρ is the density of the particle, ρ_0 is the density of ambient water, and n is the viscosity of the ambient water. Among very large particles, the maximum rates are determined to lie in the range of 1760 to 3170 m/day for freshly killed euphausiids, *Meganyctiphanes norvegica*.[108] Sinking rates of carcasses of many other zooplankters are high enough to allow settling onto the deep sea floor within several days, except when the sinking speed of dead zooplankters decreases with the reduction in size accompanying decomposition and breakage (Table 4). On the other hand, sinking rates of natural fecal pellets range from 36 to 862 m/day.[107,108] In general, the fastest sinking rates are associated with the largest fecal pellets and the slowest rates with the smallest pellets (Figure 25). These fecal pellets sink with almost the same rates as heteropods, pteropods, larger foraminifera, and amphipods, although they generally sink faster than living and dead phytoplankters. It is thus apparent that fecal pellets are as important in the vertical transport of organic materials as the carcasses of some representative zooplankters. In fact, fecal pellets are always collected in large numbers by sediment traps placed on any region of the deep sea floor.[105,109-114]

The vertical mass flux of organic materials by only these rapidly sinking large particles have been determined using sediment traps in different regions of the oceans. Vertical fluxes (mg C/m^2/day) are 7.27, 7.33, and 2.38 at the depths (km) of 1.1, 2.2, and 5.25, respectively, in the North Pacific Ocean (47° 51.1′ N, 176° 20.6′ E);[114] 6.8, 4.0, 1.7, and 1.7 at the depths (km) of 0.4, 1.0, 3.7, and 5.0, respectively, in the tropical region of the Pacific Ocean;[115] and 2.6, 0.9, and 0.7 at depths (km) of 1.0, 3.7, and 5.6, respectively, in the Sargasso Sea.[115] These determinations show that by this vertical mass flux approximately 5% of primary production in the surface layers is transported through the upper boundary of the deep water mass. This estimation is comparable to the analysis by Riley[116] that 90% of the organic matter produced by phytoplankton is utilized in the superficial layer of approximately 200 m depth since oxygen consumption in the deep water appears to be about 10% of the surface production. Rapidly sinking large particles are calculated to decompose at the rate of 1.16 μg C/m^3/day or 4.89 mg C/m^2/day as they settle through the deep water in the North Pacific Ocean from a depth of 1.1 to 5.25 km.[114] This decomposition rate is within the range of 2.3 to 14 mg $C/m^2/day$, the rates estimated for the decomposition of total organic materials in the deep water of the North Pacific Ocean. The decomposition rates of total organic materials can be estimated from the oxygen consumption rates (0.6 to 4 $\mu l/l/$ year in the deep water) by applying the Redfield-Ketchum-Richards model for organic material in seawater (C:O = 106:276 on an atomic basis). This calculation supports McCave's assumption[106] that rapidly sinking large particles contribute most of the vertical flux, while smaller particles contribute most to the concentration.

This efficient mass flux of organic materials by rapidly sinking large particles makes the relative abundance of deep-sea deposit feeders and suspension feeders dependent on the primary productivity of the overlying surface layer (Figure 26). Eutrophic environments favor the predominance of deposit feeders in nutrient-rich sediments. Oligotrophic regions in the central areas of the three oceans have low concentrations of more refractory sedimentary organic materials, resulting in the predominance of suspension feeders on the deep-sea floor.

In the oceans of the world, waters of lower latitudes are generally poorer in plankton and benthos than those of higher latitudes. The quantitative differences in biomass within the tropical regions are less pronounced than those within the subarctic or temperate regions. Decreases in biomass also are found in regions remote from the coasts. Further, decreases in biomass with depth are somewhat more rapid in lower latitudes than in higher latitudes (Figures 21 and 27). The chief factors in determining the biomass level at depth are the depth of the water column and the settling of organic materials from the photosynthetic layer, with some exceptions. The exceptions can be explained by the inflow of deep waters from highly productive regions to less productive regions this transports richer plankton biomass together with a greater amount of organic debris, both serving as food for deep-sea organisms in the lower productive regions.

Once organic materials are supplied to deep-sea organisms, the consumption of biological energy decreases markedly under higher hydrostatic pressures. This may be especially true for barophilic and eurybaric organisms. The energy consumption of barophilic organisms at *in situ* deep-sea hydrostatic conditions should be lower than that of barophobic organisms at normal atmospheric pressure. However, at equal hydrostatic pressures, higher biological activity could be exhibited by barophilic and eurybaric species. That is, deep sea organisms may show rates of activity comparable to, or higher than, barophobic species when subjected to lower pressures. The theoretical basis for this reaction is fundamentally the same as for thermal sensitivity responses of different groups, which result in the classifications of psychrophilic, mesophilic and thermophilic. Accordingly, at a given depth, organisms which have migrated upwards from deeper zones may consume more energy than those which have moved downward from shallower layers. The whole ocean, therefore, is able to support a greater biomass due to higher hydrostatic pressures in the nonphotosynthetic layers causing lowered consumption of the biological energy that is produced in the thin euphotic zone.

VII. ORGANIC NUTRIENTS

A wide variety of organic compounds is present in most aquatic environments. The major fraction of these compounds is in dissolved form at concentrations between 0.4 and 2 mg/l of water. Generally, higher concentrations are measured in the eutrophic water masses of the coastal regions, whereas the lowest are measured in the ultraoligotrophic water masses of the deep oceans.

Major chemical components of the organic solutes have the following concentrations (in mg/l): amino acids and proteins, usually less than 0.1; carbohydrates, 0.1 to 0.3;

lipids, 0.1 to 0.2. Certain minor components such as organic acids, hydrocarbons, vitamins, and hormones also are detected in natural waters. The concentrations of these compounds may be in the order of ng/ℓ, but they still can be assimilated by many organisms. The concentrations of major components in natural waters, on the other hand, are maintained in a steady-state equilibrium in the order of μg/ℓ, usually below 20 μg/ℓ. Dissolved organic materials in aquatic ecosystems thus occur not only in very low concentrations, but an appreciable fraction is tied up in biochemically resistant, humus-like materials. Therefore, only 10% or less of the organically combined materials may be readily available as substrates for aquatic microorganisms.[28]

Humic substances in natural waters are acidic polymers, which are relatively stable against biochemical degradations.[119] Humus is partly derived autochthonously from byproducts of microbial transformations of a great variety of biochemical compounds, derived from the dead cells and extracellular products of various aquatic organisms. Some aquatic humic materials are produced autochthonously from algal proteins, carbohydrates, and lipids. Phenols also are contained in marine humus formed from extracellular products of seaweeds. Other parts of the humus in natural waters and sediments may be transported allochthonously from the terrestrial environment to lakes and oceans through streams and groundwater; thus the highest proportions of allochthonous humic materials occur in certain inland waters and the coastal regions of lakes and oceans. The major part of terrestrial humus is composed of degraded lignin derived from vascular plants. These humic substances form complexes with a variety of nonhumic organic compounds, including amino acids, carbohydrates, fatty acids, phenols, and porphyrins. These metabolizable biochemical compounds, bound with humus, may be protected to a certain extent from microbial decomposition.[120] Characteristics of humus thus vary, depending primarily on the nature of the source materials. Some environmental factors also have been shown to impose additional characterization. Humification occurs both under aerobic and anaerobic conditions, although the process is especially intense under aerobic conditions. In the particular case of aquatic humus, the degree of humification is influenced by sedimentation; humus in deep sediments is slightly more condensed than in shallow sediments.

Aquatic humus precipitates through the processes of flocculation, adsorption to clay particles, incorporation into the fecal pellets of filter-feeding invertebrates, aggregation on bubbles, and by hydraulic factors that regulate the mechanical deposition of sediments. Sedimentary humus is especially concentrated in clays and silts in coastal regions of the marine environment as well as in freshwater lakes and rivers. The adsorption capacity of the clay mineral depends upon the species of mineral and humic compounds.

Humus may have harmful effects on aquatic ecosystems, as the low productivity of certain humus-rich environments may be caused by immobilization of micronutrients, absorption of light, excessive acidity, and possibly the presence of antibiotic substances such as phenols. However, the beneficial effects of aquatic humus must be greater. One of the beneficial functions has been ascribed largely to its metal-binding and cation-exchange properties; humus aids detoxification by the scavenging of heavy metals and other dissolved poisonous substances. Humus also makes nutrient cations such as Fe^{3+} available to algae, enhances the respiration and nutrient uptake of microbial cells, performs a function in microbial physiology, and stimulates microbial growth. The binding and release of phosphate by metal humates could be vitally important to aquatic organisms. Above all, the most beneficial effect of aquatic humus must be in the contribution to the stability of aquatic ecosystems, since humus is an important reservoir of biological elements. Aquatic humic compounds are not completely refractory to biodegradation basically because they are synthesized by biochemical reactions. The humification processes cause the biological elements of biochemical compounds to be converted partially into relatively refractory substances instead of keeping them in easily metabolizable states; thus humification tends to slow the average recycling rate of each biological element, leading to increased stability of organic materials in aquatic ecosystems.

A predominant type of heterotrophic bacteria in natural waters is free-living bacterioplankton that must depend almost exclusively on obtaining nutrient requirements from dilute nutrient solutions. In natural waters, microbial assimilation of a limiting nutrient is assumed to be performed by the interactions of various transport systems among microbial species having different metabolic pathways. These complicating interactions in natural populations cannot be simulated in any experimental laboratory condition. However, using a chemostat to produce experimental steady-state conditions, these difficulties are eliminated to a certain degree by the use of time-independent relationships between culture conditions and growth properties of a microorganism. It is then possible to assess its growth responses to extremely low concentrations of the limiting substrate in the medium.[121] The threshold concentrations for organic nutrients and the maximum growth rates differ even among strains of microorganisms all isolated from the marine environment (Table 5).

Pseudomonas sp. (Strain 201), *Spirillum* sp. (Strain 101), and *Achromobacter* sp. (Strain 317) clearly represent inefficient species; they are unable to become free-living bacterioplankton at the low substrate concentrations of natural seawater as they require higher concentrations of desirable nutrients than the concentration of total organic compounds in the seawater. Thus they may not exist exclusively in the form of free-living bacterioplankton in the marine environment. On the other hand, the threshold concentrations for growth of efficient species (Figure 28) such as *Achromobacter aquamarinus* (Strain 208) and *Spirillum lunatum* (Strain 102), still appear to be more than ten times higher than the concentration of each organic compound in natural waters (i.e., usually below 20 μg/ℓ). These threshold concentrations of efficient species, however, are almost the same level or a little less than the concentrations of total dissolved organic materials in natural waters (i.e., 400 to 2000 μg/ℓ). Under natural conditions, moreover, heterotrophic microorganisms with these nutrient requirements could assimilate various substrates more efficiently in the presence of a variety of species with increased metabolic diversities and varieties of growing constants. Heterotrophic bacteria belonging to these efficient species can be assumed to exist as free-living bacterioplankters, completely adapted to the dilute nutrient environments of oligotrophic water masses. As the concentration of organic solutes in natural water should be maintained at threshold levels by the active transport systems of assimilation by natural bacterial populations, the relatively high concentration of organic materials actually dissolved in seawater is due to the appreciable fractions of materials with relatively high biochemical resistance against microbial decomposition. *Vibrio* sp. (Strain 204) is an intermediate between inefficient and efficient species. Such intermediate species possibly have more efficient active transport systems of nutrient assimilation for some other substrates in natural waters.

Another predominant type of heterotrophic bacteria in natural waters consists of stalked forms, adherent to the interface between a liquid phase and a solid or gaseous phase. Stalked bacteria commonly are found in aquatic environments attached to particulate matter, either plant or animal materials. Most members of this type belong to the genus *Caulobacter*. The cells of *Caulobacter* often occur in rosettes which adhere to one another by means of a common mass of hold-fast material. *Caulobacter* species are heterotrophic aerobes that are able to grow on a variety of dilute organic nutrients. They have a special process of cell division with unequal binary fission. Two different kinds of cells occur in any *Caulobacter* species (Figure 29): swarmer cells and stalked cells. The swarmer bears a single polar flagellum and is grossly indistinguishable from a simple motile unicellular bacterium. The second kind of cell bears a stalk at one pole in place of a flagellum and is thereby nonmotile. The stalk formation is an obligatory stage in the development of every swarmer, attached or unattached, and always precedes the first division of the swarmer. When a stalked cell divides into two dissimilar sister cells, division of the stalked sibling always precedes division of the swarmer sibling. However, after the swarmer has formed a stalk and divided once, its subsequent division times are comparable with those of its original stalked sibling.[123]

Caulobacter has the ability to grow in waters which are very low in organic matter.[123,124] This ability is only possible however, when *Caulobacter* can find a microenvironment high in organic concentration. The cell division cycle of *Caulobacter* favors location of a microenvironment satisfying nutrient requirements in the following manner: a swarmer cell separates from the stalked mother cell, swims until it locates a suitable microenvironment, and then settles down on the new surface by forming a new stalk. Thus, in oligotrophic environments, the *Caulobacter* cell can acquire greater amounts of the restricted supply of organic nutrients than can other microorganisms which lack this appendage.

There must be many places where even stalked bacteria, as well as bacterioplankton, find it difficult to satisfy their nutritional requirements. This may be true of many locations in the deep oceans. In such environments, the mixotrophic mode should become prevalent, because mixotrophic microorganisms can assimilate organic compounds as sources of materials while using inorganic compounds as energy sources. Mixotrophs thus grow best among heterotrophic communities in ultraoligotrophic conditions.

Facultative autotrophic bacteria, such as *Thiobacillus novellus* and *T. intermedius*, may be representatives among mixotrophs in aquatic environments. These sulfur-metabolizing bacteria may play a significant role in the transformations of organic and inorganic materials in the marine environment because sulfur is one of the most abundant anions in seawater. These facultative autotrophic thiobacilli are quite different from the strictly autotrophic thiobacilli such as *Thiobacillus thioparus, T. thiooxidans,* and *T. denitrificans.* The gap between the strictly and facultatively autotrophs is bridged by some strictly autotrophic thiobacilli, such as *Thiobacillus neopolitanus* and *T. ferroxidans,* which can adapt to assimilate certain organic materials in the presence of an oxidizable inorganic sulfur compound.

The facultative autotrophic bacteria, *Thiobacillus novellus* and *T. intermedius*, are presumed to be distributed widely in aquatic environments. *T. intermedius* was first isolated from freshwater mud.[125] This strictly aerobic bacterium is supported by either thiosulfate or organic compounds, and its optimal growth occurs in the presence of glutamic acid and glucose. *T. novellus* was first isolated from soil,[126] and its marine strain has been isolated from a shallow estuary of Kaneohe Bay, Oahu Island, Hawaii.[7] This bacterium is a Gram-negative, strictly aerobic, nonmotile, nonsporulating rod (0.5×1.5 μm), with highly pleomorphic forms in the heterotrophic growth phase (Figure 30). This bacterium can oxidize thiosulfate, elemental sulfur, sulfite, and tetrathionate to generate energy which may be utilized for carbon dioxide fixation (Table 6). Sulfite oxidation cannot provide enough energy for carbon dioxide fixation and the energy generated by the oxidations of sulfur and tetrathionate also is too small to support autotrophic life. Thiosulfate seems to be the only substrate which produces enough energy to support autotrophic life. Autotrophically growing cells of the bacterium are converted easily to excellent heterotrophic growth in the presence of glucose, galactose, glycerol, mannose, fumarate, succinate, and lactose as well as in the presence of yeast extract, peptone, or beef extract. As autotrophic growth with a thiosulfate-oxidizing system is produced in heterotrophically grown cells only when thiosulfate is present in the growth medium, this system is believed to be inducible. Cells with the thiosulfate-oxidizing system can oxidize thiosulfate more rapidly than galactose (Figure 31). Both galactose and thiosulfate can be utilized by the cells at the same time, but more oxygen is required for the oxidation of both substrates together than for the oxidation of each substrate separately. Another bacterial strain, *Thiobacillus* A2, that shares common habitats with *T. novellus*,[127] has the ability to grow heterotrophically at faster rates and on a greater range of organic compounds (Table 7).

These facultative autotrophic thiobacilli have been shown to be an appreciable fraction of marine microbial populations, forming more than 25% of bacterial populations that can be cultured from water and sediment samples of various oceanic regions on common heterotrophic medium.[128] On the other hand, strict autotrophic thiobacilli are rare in the marine environment. Hence, biological oxidation of reduced sulfur compounds, in waters from the surface to the great depths of the oceans, is estimated to be carried out largely by facultative autotrophic thiobacilli.[129]

Although these mixotrophic bacteria, such as strains of the species *Thiobacillus novellus* and *T. intermedius*, are all regarded as facultatively autotrophic, some obligately mixotrophic bacteria are found in nature. *T. perometabolis* is a typical obligate mixotroph which is not capable of strict autotrophic growth.[130] Its growth on organic compounds only is limited in rate and yield. Optimal growth conditions require the presence of a partially reduced sulfur compound and an organic compound such as yeast extract, fructose, arabinose, ribose, or xylose. All types of mixotrophic thiobacilli can be assumed to be widespread throughout the marine environment based on the evidence that thiosulfate exists not only at oxygen-sulfide interfaces of such specific locations as in the Black Sea, Saanich Inlet, and the Cariaco Trench, but also in the surface layer of the open ocean.[131]

The upper limit of the concentration of dissolved organic materials in oligotrophic waters is maintained by aquatic bacteria at a threshold level usually below 1 mg/ℓ. Therefore, it is much easier for microorganisms having thresholds higher than 1 mg/ℓ to live at an interface in natural waters, preferably with one side of the interface being solid.[10] This is chiefly because solid surfaces promote growth of the attached microorganisms by providing a higher concentration of organic nutrients than that present in the surrounding water. The chances for particulate organic materials to be adsorbed on the substratum increase with greater velocity up to a certain limit, beyond which the water flow may prevent their adsorption or even dislodge them from solid surfaces. The adsorption of microorganisms on a submerged substratum also is influenced by the texture or smoothness of the substratum, its cleanliness, chemical composition, ion-exchange properties, hardness or cohesiveness, its shape and area, surface tension, pH, electrostatic properties, and certain other characteristics or conditions.[132] Once the microorganisms attach to the substratum, the solid surface influences the metabolic and physiological activities of the organisms by way of:

1. Concentration of organic nutrients on the substratum
2. Orientation of large polar molecules on solid surfaces
3. Minimization of the diffusion of exoenzymes, hydrolyzates, or metabolites away from the organisms
4. Neutralization of the zeta potential or other electrostatic conditions on the cell wall, thereby influencing permeability and leakage phenomena
5. Creation of microenvironmental conditions which may be quite unlike conditions in the surrounding water[132]

ZoBell and Anderson[10] have shown experimentally that marine bacteria flourish best when provided with the greatest solid surface area per unit volume of seawater. The efficiency of their activities also is shown to be enhanced due to adsorption of organic materials on solid surfaces and to the reduction of diffusion at the immediate surfaces of substratum, especially within the interstices at the tangent of the bacterial cell and the solid surfaces. Oxygen consumption does not continue logarithmically until all the oxygen is depleted, but it starts to retard between the 15th and 21st days of incubation (Figure 32). Oxygen consumption becomes very slow after 28 days storage at $16°C$. This general trend on the oxygen consumption of 1000 mℓ samples stored in 1000 mℓ glass-stoppered bottles continues at a slower rate and for longer periods of time than that of 100 mℓ samples stored in 100 mℓ glass-stoppered bottles. This favorable influence of small volumes on the activity and multiplication of bacteria is attributed to the contact of water with the larger solid surface area in the small receptacles. Regardless of the water volume of the sample, oxygen is not depleted from the seawater; there are still between 2 and 3 mℓ of oxygen per liter when oxygen consumption virtually ceases. This means that organic materials in the seawater sample can be calculated, using the Redfield-Ketchum-Richards model (C:O = 106:276 on the atomic basis), to decrease down to approximately 0.35 mg/ℓ by microbial decomposition, providing that the concentration of organic materials in the original seawater sample was 1 mg/ℓ (or 0.4 mg C/ℓ) which is an average value in the oceans.[38] This volume effect has been shown to disappear when more than a few mg/ℓ of utilizable organic materials are added to the seawater. Therefore, the favorable influence of solid surfaces on microbial activity and growth is shown, at least with organic concentrations ranging between 0.35 mg to a few mg/ℓ. As this experimentally determined range of organic concentrations corresponds exactly to those actually measured in every location of the hydrosphere, the favorable influence of solid surfaces on the decomposition of organic materials can be especially important in regulating the lower level of organic materials in natural waters.

The inorganic fraction of particulate matter in the marine environment amounts to at least 70%. Calcium carbonate particles are known to occur predominantly in the open ocean, whereas inorganic particles associated with clay and other minerals derived from the terrestrial region predominate in coastal areas. These minerals include appreciable amounts of silicon, iron, aluminum, and calcium. These inorganic particles provide desirable solid surfaces for adsorption of organic materials, as has been shown experimentally by Bader et al.[133] An absorbed layer of organic materials on carbonate particles in tropical and subtropical seas also has been demonstrated.[134] Interaction of dissolved organic materials with inorganic particles leads not only to adsorption but also to polymerization of organic materials. The surfaces of clay minerals and silicate are known to be effective polymerization agents for organic monomers, and they also are catalysts.[135] In consequence, biochemical compounds are believed to be condensed and also chemically altered by means of inorganic-organic interactions on particle surfaces. These chemical alterations could favor the transformation of biochemically stable substrates into metabolizable materials for aquatic microorganisms.

The conversion of soluble organic materials into particulate form may be performed by the accumulation of organic solutes at the interface between liquid and gaseous phases. Sutcliffe et al.[136] showed that when air is bubbled through filtered seawater, organic particles are found in the spray droplets formed from bursting bubbles. The bubbling is performed very easily by wave action in the surface layer of the hydrosphere. Carlucci and Williams[137] showed that the action of bubbling seawater tends to concentrate bacteria in the foam. When large hydrophilic molecules adsorb to bubbles, monomolecular films are produced. The films may aggregate into insoluble organic particles during the aeration process to produce foam; the ever increasing area of the adsorbed monomolecular films causes them to fold into polymolecular layers and to form colloidal micellae or to collapse into fibers. Then the agitation of foaming produces repeated collisions and results in multiple-layer coalescence of colloidal particles to produce a semistable suspension of organic materials. Finally, many of these larger particles undergo further aggregation.[79] Many of the marine aggregates appear as long streaks *in situ* and are known as "marine snow".

Bacteria may play some role in the formation of aggregates as suggested by Barber.[138] One of the most important characteristics of aquatic bacteria is the tendency for clumping. Many of them have pili or excrete polymers capable of cementing action so that these bacteria can stick together, forming larger particles with inclusion of other materials.[32] A large particle, like an aggregate is a nutrient-rich microenvironment and therefore, the bacteria which contributed to its formation can continue to live inside the aggregate. In this favorable environment even such a bacterium as *Escherichia coli*, which should be nutritionally inactive in natural waters, can survive as well as some aquatic bacteria which are not completely adapted to the bacterioplanktonic life by their inability to grow at naturally low substrate concentrations. Incidentally, *E. coli* is a very inefficient species, requiring high threshold concentrations of nutrients (e.g., 10 mg/ℓ peptone).

At every location in the hydrosphere, aggregates containing bacteria and phytoplankters can be detected easily (Figure 33). Actually, many kinds of phytoplankters are known which can survive for long periods in complete darkness, some of them up to a year.[139,140] These phytoplankters living in an aggregate should have some meaningful role in the dynamics of organic materials in aquatic environments.

Most aquatic environments viewed in macroscopic dimension appear to be very dilute nutrient environments. However, viewed more closely, such environments are shown to contain scattered nutrient-rich microenvironments, where it is even possible for heterotrophic microorganisms with low efficiencies of nutrient assimilation to survive.

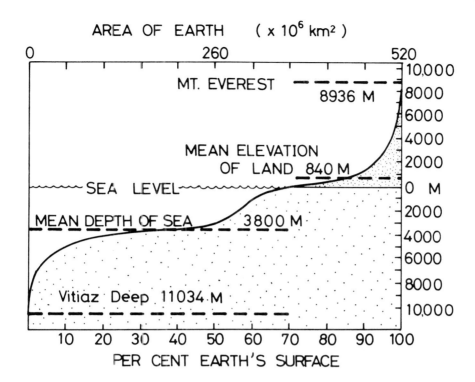

FIGURE 1. Relative areas and elevations of land and water masses on the surface of the Earth. Whereas approximately 72% of the surface of the Earth is covered with water to an average depth of 3800 m, the mean elevation of land is 840 m. The greatest known depth and highest elevation are also shown. (From ZoBell, C. E., *Low Temperature Microbiology Symposium*, Campbell Soup Co., Camden, N.J., 1961, 107. With permission.)

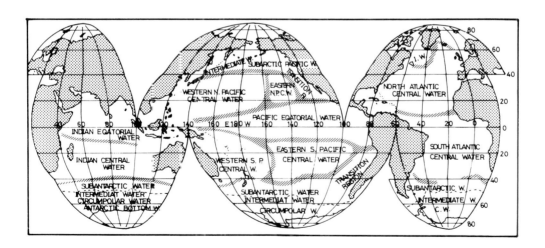

FIGURE 2. Approximate boundaries of the upper water masses of the world ocean. (From Sverdrup, H. U., Johnson, M. W., and Fleming, R. H., *The Oceans. Their Physics, Chemistry, and General Biology*, Prentice-Hall, Englewood Cliffs, N.J., 1942, 1087. With permission.)

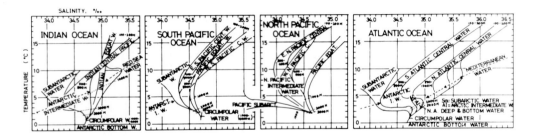

FIGURE 3.   Temperature-salinity relations of the principal water masses of the oceans. (From Sverdrup, H. U., Johnson, M. W., and Fleming, R. H., *The Oceans. Their Physics, Chemistry, and General Biology,* Prentice-Hall, Englewood Cliffs, N.J., 1942, 1087. With permission).

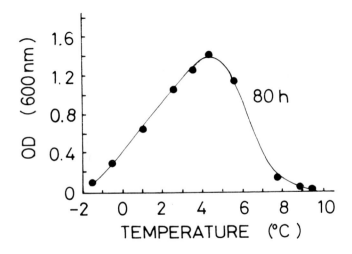

FIGURE 4.   Growth of an Antarctic psychrophile (AP-2-24, designated tentatively as a *Vibrio* sp.). Incubation period was 80 hr in Lib-X medium employing a temperature gradient incubator. Lib-X medium contains: yeast extract (Difco), 1.2 g; Trypticase (BBL), 2.3 g; sodium citrate, 0.3 g; L-glutamic acid, 0.3 g; sodium nitrate, 0.05 g; ferrous sulfate, 0.005 g; Rila marine mix, 33 g; and distilled water, 1000 mℓ. The pH was adjusted to 7.5. (From Morita, R. Y., *Bacteriol. Rev.,* 39, 144, 1975. With permission.)

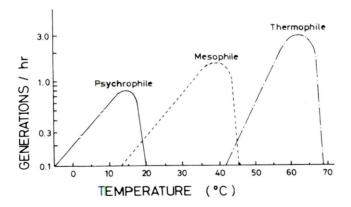

FIGURE 5. Relation of temperature to growth rates of a psychrophile, a mesophile, and a thermophile. (From Brock, T. D., *Biology
of Microorganisms,* Prentice-Hall, Englewood Cliffs, N.J., 1979, 802.
With permission.)

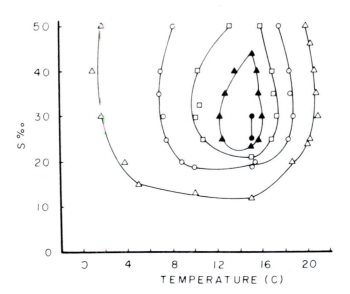

FIGURE 6. Contour surface diagram showing the coeffect of temperature and salinity on the growth rate as a number of generations/
hr of *Vibrio marinus* MP-1 in Lib-X broth. Symbols: •, 0.28 genera-
tions/hr, (maximum); ▲, 0.25 generations/hr; □, 0.20 generations/
hr; ○, 0.14 generations/hr; and △, 0.06 generations/hr. (From Morita, R. Y., *Bacteriol. Rev.*, 39, 144, 1975. With permission.)

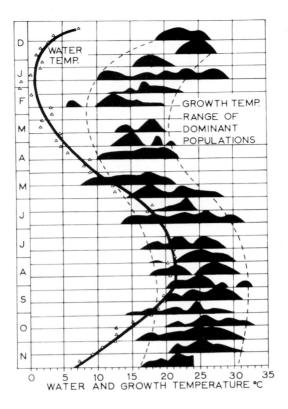

FIGURE 7.   Comparison of the water temperature
curve with the range of maximal growth for the dominant microflora from December 1960 to November
1961, in Narragansett Bay, Rhode Island. (From Sieburth, J. McN., *J. Exp. Mar. Biol. Ecol.,* 1, 98, 1967.
With permission.)

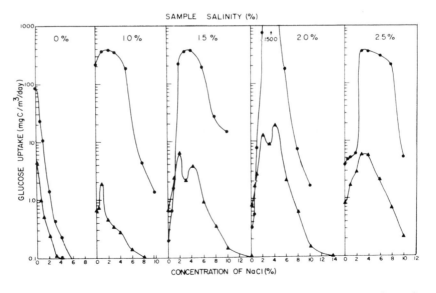

FIGURE 8.   Heterotrophic activity of water samples at different concentrations of
NaCl (•, data from July; ▲, data from November). (From Seki, H., Stephens, K. V.,
and Parsons, T. R., *Arch. Hydrobiol.,* 66, 37, 1969. With permission.)

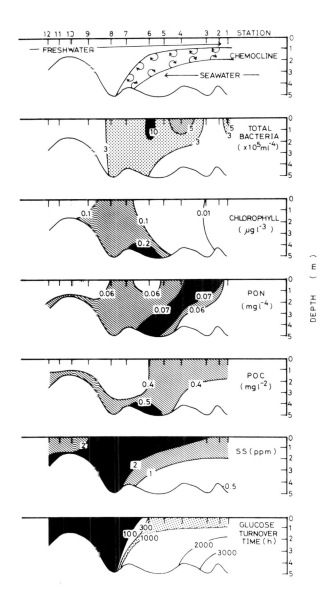

FIGURE 9. Environmental and microbial transitions in the lower reaches of the River Teshio. (From Seki, H. and Ebara, A., *J. Oceanogr. Soc. Jpn.*, 36, 30, 1980. With permission.)

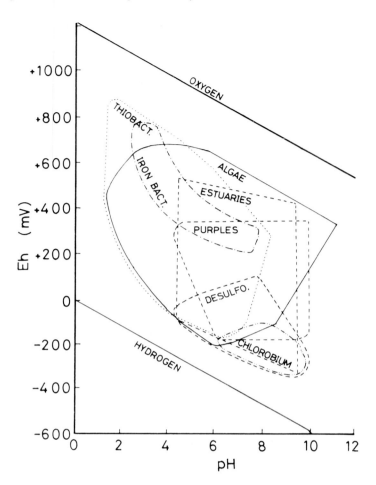

FIGURE 10.   The range of bacterial groups and algae in aqueous systems with respect to pH and Eh. (From Oppenheimer, C. H., *Geochim. Cosmochim. Acta*, 19, 244, 1960. With permission.)

FIGURE 11. Summer kill of fish and shellfish in Lake Kasumigaura, Japan. The fish and shellfish have died from lack of oxygen and are floating on the lake surface.

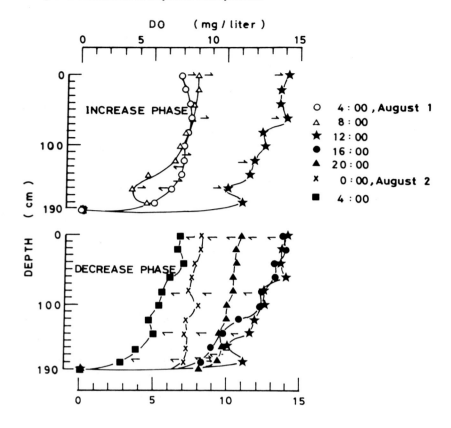

FIGURE 12.   *In situ* dissolved oxygen profiles measured at 4-hr intervals, started from 4:00 a.m. on August 1 to 4:00 a.m. on August 2, 1977. (From Seki, H., Takahashi, M., Hara, Y., and Ichimura, S., *Water Res.*, 14, 179, 1980. With permission.)

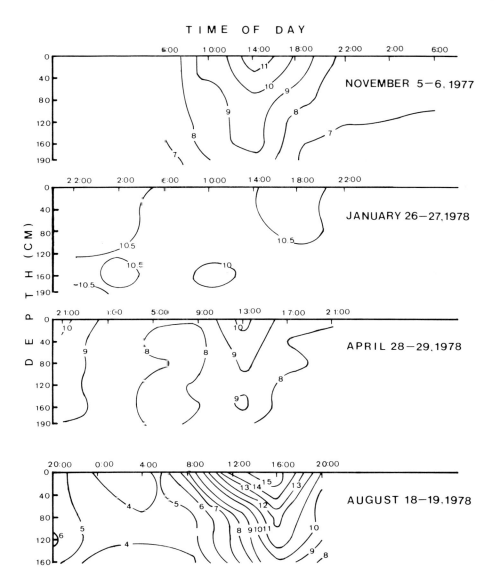

FIGURE 13. Diel fluctuation of dissolved oxygen (mg O$_2$/ℓ) at each season in Lake Kasumi-gaura, Japan. (From Kuroiwa, K., Ogawa, Y., Seki, H., and Ichimura, S., *Water Air Soil Pollut.*, 12, 255, 1979. With permission.)

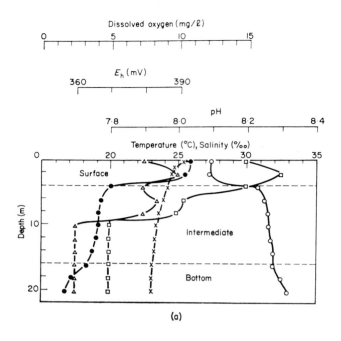

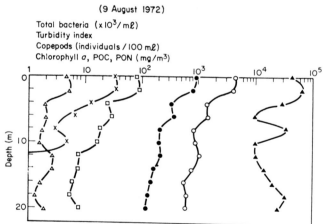

FIGURE 14.   Vertical distribution in (a) of temperature (x), salinity (o), pH (□), Eh (△), dissolved oxygen (•), and in (b) of particulate organic carbon (o), particulate organic nitrogen (•), chlorophyll *a* (□), turbidity (△), copepods (x) and total bacteria (▲) on August 9, 1972. (From Seki, H., Tsuji, T., and Hattori, A., *Est. Coast. Mar. Sci.*, 2, 145, 1974. With permission.)

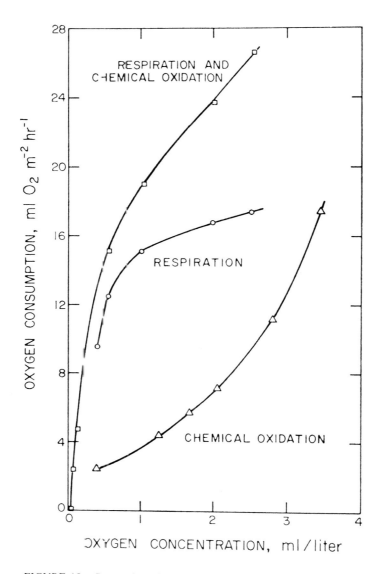

FIGURE 15.   Rates of total oxygen uptake, inorganic chemical oxidation, and respiration of a single core as functions of oxygen concentration. This core came from 74 m depth in Port Madison, Puget Sound, Washington. The experiment, conducted at 9.0°C, lasted 2 days. The total uptake was measured until little oxygen remained; then the electrode was withdrawn, the deoxygenated water was flushed out with equilibrated filtered (Millipore) water without disturbing the sediment. Formaldehyde was added, and the core was resealed. (From Pamatmat, M. M., *Limnol. Oceanogr.*, 16, 535, 1971. With permission.)

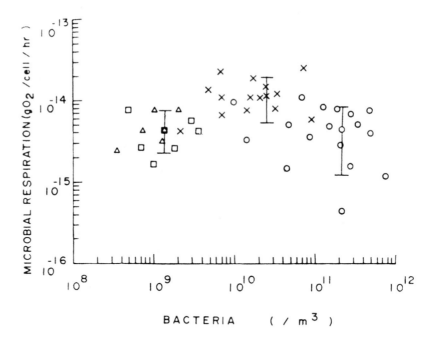

FIGURE 16.   Relationship between population density of bacterioplankton and their respiration in seawater of Tokyo Bay, Aburatsubo Inlet (x), the subarctic Pacific water (△), and the western North Pacific central water (□). (From Seki, H., *La Mer*, 11, 147, 1973. With permission.)

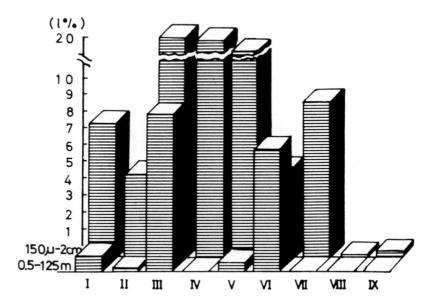

FIGURE 17.   Frequency distribution histogram of saprophytic bacterial genera of bacterioneuston and bacterioplankton in the 150μ to 2-cm layer and the 0.5 to 125-m layer of the Aleutian Gut. The frequency of a given genus (l in %) refers to the number of all the bacterial strains isolated from the Aleutian Gut. (I) *Bacillus*; (II) *Micrococcus*; (III) *Bacterium*; (IV) *Chromobacterium*; (V) *Pseudomonas*; (VI) *Mycobacterium*; (VII) *Planococcus*; (VIII) *Streptococcus*; (IX) *Pseudobacterium*.[76] (From Tsiban, A. V. and Teplinskaya, N. G., *Biological Oceanography of the Northern North Pacific Ocean*, Takenouti, A. Y., Ed., Idemitsu Shoten, Tokyo, 1972, 541. With permission.)

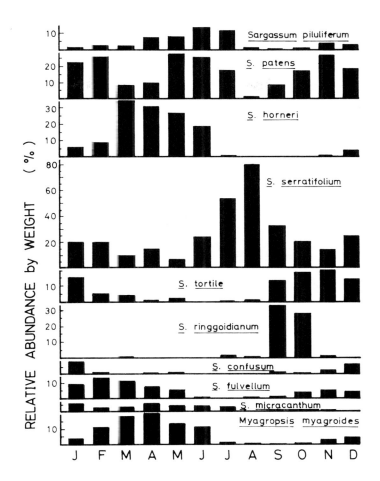

FIGURE 18. Seasonal fluctuation in relative abundance of species found in the floating algae collected off a coast in the southern part of Japan. (From Yoshida, T., *Bull. Tohoku Reg. Fish. Res. Lab.*, 23, 141, 1963. With permission.)

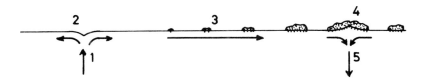

FIGURE 19. Scheme of formation of neustonic concentrations in convergence zones; (1) upwelling of biogen-rich deep waters; (2) divergence zone with rich phytoplankton and poor neuston; (3) transfer of water to the convergence zone, accompanied by gradual proliferation of the neuston; (4) convergence zone with abundant neuston, poor phytoplankton, and large amounts of foam and floats; (5) sinking of waters, primarily containing macrozooplankton. (From Zaitsev, Yu. P., *Marine Neustonology*, Israel Program for Scientific Translations, Keter Publishing, Jerusalem, 1971, 207. With permission.)

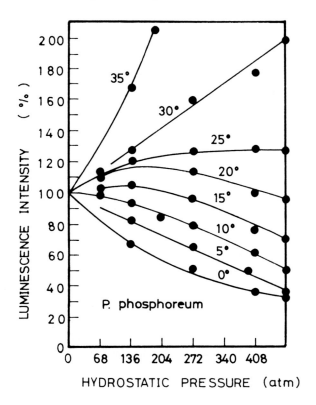

FIGURE 20. Intensity of luminescence of *Photobacterium phosphoreum* as a function of pressure at different temperatures. The intensity at normal pressure is arbitrarily taken equal to 100 at each temperature in order to show the percent change in intensity with change in pressure. (From Johnson, F. H., *The Kinetic Basis of Molecular Biology*, John Wiley & Sons, New York, 1954, 310. With permission.)

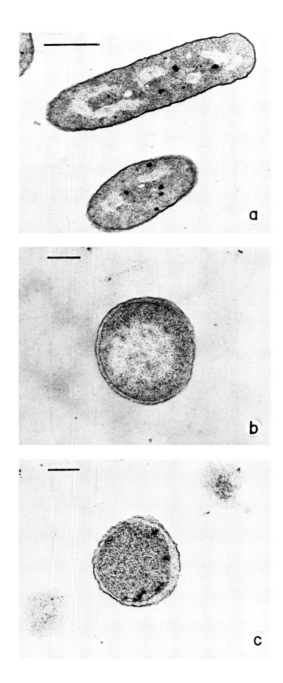

FIGURE 21. Electron micrographs of *Vibrio* Ant-300 cells: (a) nonstarved cells, bar represents 1.0 μm; (b) and (c) cells starved for 5 weeks, bar represents 0.2 μm. (From Novitsky, J. A. and Morita, R. Y., *Appl. Environ. Microbiol.*, 32, 617, 1976. With permission.)

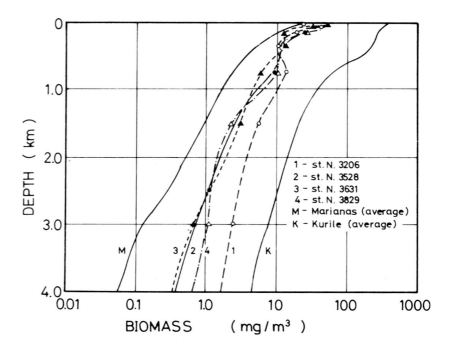

FIGURE 22.   Vertical distribution of plankton biomass at some stations in the zone of an inflow of deep waters from temperate latitudes. Distribution in the Kurile-Kamchatka trench (K) and tropical Marianas trench (M) given for comparison. (From Vinogradov, M. E., *Deep Sea Res.*, 8, 251, 1962; copyright Pergamon Press. With permission.)

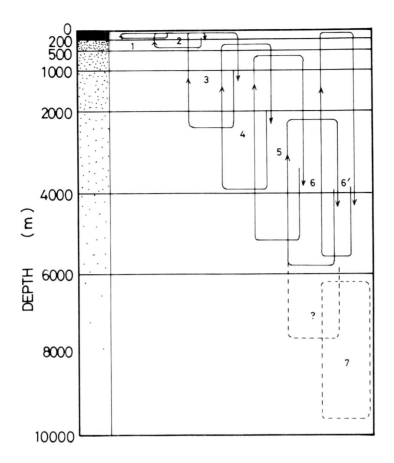

FIGURE 23. Scheme of vertical migrations of the deep-sea plankton. (1) Migrations of the surface species; (2) migrations extending over the surface and transition zone; (3) migrations extending over the surface, transition, and upper layers of deep-sea zones; (4) migrations extending over transition and part of the deep-sea zone; (5) migrations within the whole deep-sea zone; (6) irregular migrations of some species extending through the whole water column; (7) range of distribution of ultra-abyssal animals. Variations of the plankton abundance with increasing depth are shown by frequency of dots in each layer that is proportional to the biomass of the plankton.[101] (From Vinogradov, M. E., *Rapp. Proc. Verb. Cons. Perm. Int. Explor. Mer.*, 153, 114, 1962. With permission.)

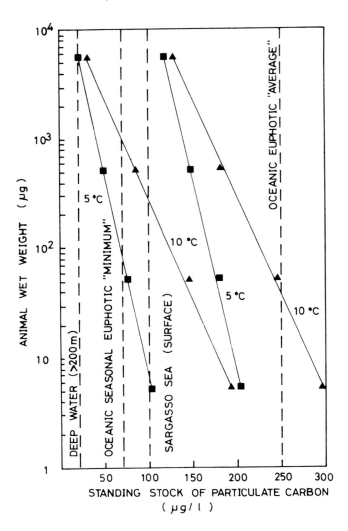

FIGURE 24.   Approximate standing stock of particulate carbon required by different-sized zooplankton at 5° and 10°C and at two growth rates (0.7%, ■—■ ; and 7%, ▲—▲ per day). (From Parsons, T. R. and Seki, H., *Organic Matter in Natural Waters* Hood, D. W., Ed., University of Alaska, Fairbanks, 1970, 1. With permission.)

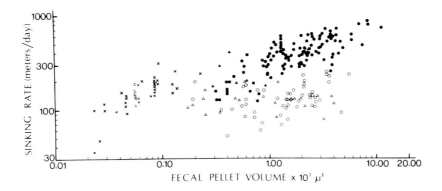

FIGURE 25.   The rela:ionship between sinking rate and volume for natural and labo-
ratory-produced eμphausiid fecal pellets. Fecal pellets originated from *Meganyctiphanes norvegica*, r atural food (•); *Euphausia krohnii*, natural food (▲); *Nematoscelis
megalops*, natural :ood (■); *M. norvegica*, fed *Artemia* that had ingested flagellates
(o); *M. norvegica* a nd E. *krohnii*, fed *Phaeodactylum tricornutum* (△). Fecal pellets of
unknown origin from the data of Smayda[107] are plotted for comparison (x). (From
Fowler, S. W. and Small, L. F., *Limnol. Oceanogr.*, 17, 293, 1972. With permission.)

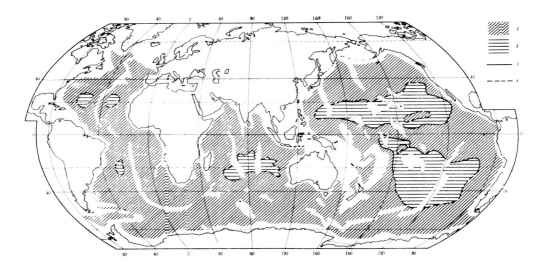

FIGURE 26.   Deep-sea trophi: regions of the ocean bottom (depths greater than 3000 m). (1) Eutrophic;
(2) oligotrophic; (3, 4) boundaries between regions. (3) Plot according to weight predominance of deposit-
feeders or suspension-feeders in trawl hauls. (4) Plot according to distribution of major taxonomical groups
of macrobenthic detritus-feeders and to conditions of organic matter accumulation and its transformation
in surface layer of bottom sediments. (From Sokolova, M. N., *Mar. Biol.*, 16, 1, 1972. With permission.)

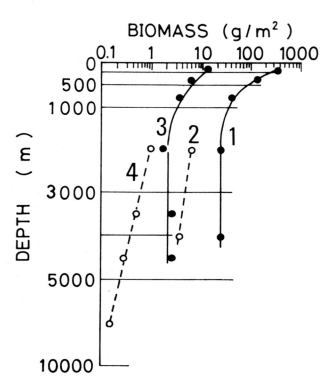

FIGURE 27.   Change in the biomass of the benthos with depth
(1) in the 100-mile coastal zone in the northwestern part of the
Bering Sea; (2) at a greater distance from the shore in the western
half of the Bering Sea; (3) in the 100-mile coastal zone of the tropical part of the Indian Ocean; (4) in deep water at a greater distance from the shore in the tropical part of the Indian Ocean.
(From Belyayev, G. M. and Vinogradova, N. G., *Oceanology*,
136, 35, 1961. With permission.)

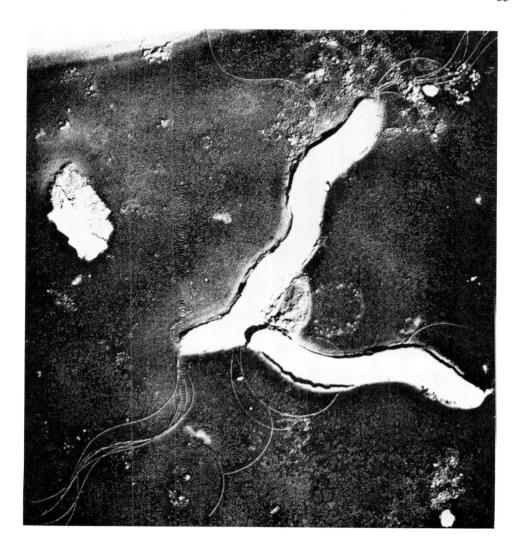

FIGURE 28. *Spirillum lunatum* Strain 102 (presently identified as a strain of *Oceanospirillum maris*[122]) adapted to the marine environment by its ability to grow at low substrate concentration. (From Hylemon, P. B., Wells, J. S., Jr., Krieg, N. R., and Jannasch, H. W., *Int. J. Syst. Bacteriol.*, 23, 4, 1973. With permission.)

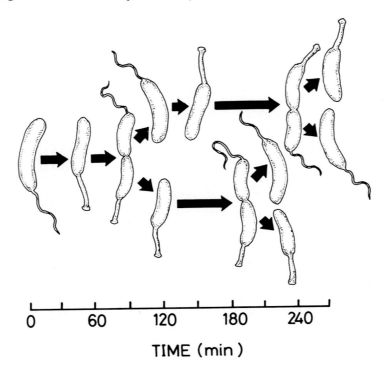

TIME (min)

FIGURE 29.   The cell cycle of *Caulobacter*. (From Stove, J. L. and Stanier, R. Y., *Nature (London)*, 196, 1189, 1962. With permission.)

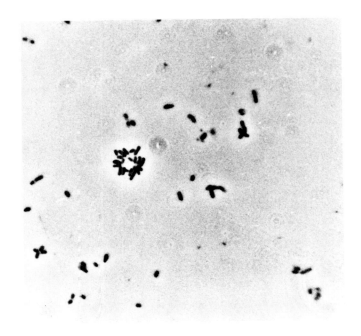

A

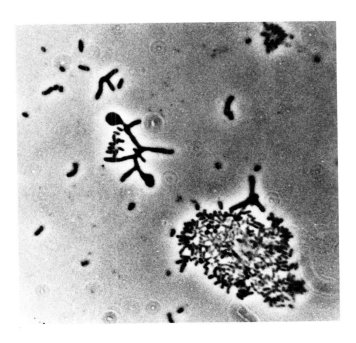

B

FIGURE 30. Phase photomicrographs of normal cells (A) and pleomorphic cells (B) of strain D10-B, a facultative chemoautotrophic sulfur bacterium. (From Adair, F. W. and Gundersen, K., *Can. J. Microbiol.*, 15, 355, 1969. With permission.)

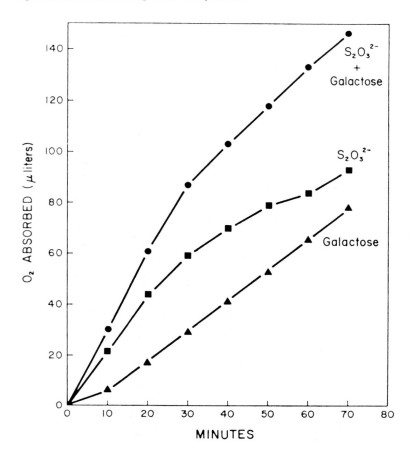

FIGURE 31.   Oxidation of galactose and thiosulfate by galactose-thiosulfate-grown cells. (From Adair, F. W. and Gundersen, K., *Can. J. Microbiol.*, 15, 355, 1969. With permission.)

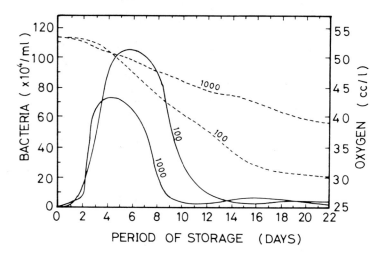

FIGURE 32.   Influence of volume on oxygen consumption in stored seawater. Solid lines represent growth curves and dashed lines oxygen consumption in 100 and 1000 ml seawater volumes, respectively. (From ZoBell, C. E. and Anderson, D. Q., *Biol. Bull.*, 71, 324, 1936. With permission.)

FIGURE 33. Aggregate with the appearance of long streaks collected at a depth of 4000 m in the region of the western North Pacific central water (28° 25′ N, 145° E). Many bacteria and phytoplankton fragments are adherent.

Table 1

PHYSICAL MIXING PARAMETERS AND OXYGEN
CONSUMPTION RATES IN THE DEEP WATER OF THE NORTH
PACIFIC

Location				Oxygen consumption rates[a]		
Lat.	Long.	W/D km^{-1}	α km^{-1}	at 3 km, ml/m^3/year	below 1 km, l/m^2/year	below 1.5 km, l/m^2/year
51°N,	170°W	1.1	1.0	1.5	11.0	6.5
47°N,	146°W	1.2	2.0	1.0	26.0	9.6
41°N,	146°W	1.3	1.5	1.4	18.6	8.7
40°N,	158°E	1.1	1.5	1.3	16.7	7.9
39°N,	170°W	1.2	1.5	1.7	22.5	10.6
32°N,	146°W	1.1	1.5	1.3	17.6	8.3
30°N,	170°W	1.2	1.5	1.0	13.1	6.2
27°N,	158°E	1.3	0.4	1.6	6.6	5.0
24°N,	170°W	1.0	0.3	1.4	5.4	4.2
20°N,	146°W	1.0	0.3	2.2	8.6	6.7
18°N,	158°E	1.0	0.3	2.1	8.1	6.4
15°N,	170°W	0.8	0.2	1.3	5.0	4.0
12°N,	158°E	0.8	0.2	1.6	5.9	4.8
6°N,	170°W	0.7	0.2	1.4	5.4	4.4
3°S,	170°W	0.7	0.2	1.3	4.9	4.3

[a]   In the calculation, $D = 1.3$ cm^2/sec was used.

From Tsunogai, S., *Biological Oceanography of the Northern North Pacific Ocean,* Takenouti, A. Y., Ed., Idemitsu Shoten, Tokyo, 1972, 517. With permission.

Table 2

RELATIVE TURBIDITY CAUSED BY THE MULTIPLICATION OF MARINE BACTERIA IN NUTRIENT BROTH FOR 6 DAYS AT 20°C, 4 DAYS AT 30°C, OR 1 DAY AT 40°C AT DIFFERENT HYDROSTATIC PRESSURES

Culture	300 Atm			400 Atm			500 Atm			600 Atm		
	20°C	30°C	40°C	20°C	30°C	40°C	20°C	30°C	40°C	20°C	30°C	40°C
Achromobacter fischeri	+++ +	++	A	++	−	−	−	−	−	−	−	−
Achromobacter harveyi	+++ +	+++ +	++	+	+++ +	−	−	−	−	−	−	−
Achromobacter thalassius	−	+++ +	++	−	−	+++	−	−	−	−	−	−
Bacillus abysseus	+	+++ +	+++ +	−	+++ +	+++ +	−	+	+++ +	−	−	+++ +
Bacillus borborokoites	++	++++ +	+++ +	−	++++ +	++++ +	−	++	+++ +	−	−	+++ +
Bacillus cirroflagellosus	++	+++	A	−	++		−	−		−	−	
Bacillus submarinus	++	++++ +	+++ +	+	++++ +	+++ +	−	+++ +	+++ +	−	+++ +	+++ +
Bacillus thalassokoites	+++	++++ +	+++ +	++	++++ +	+++ +	−	+++ +	+++ +	−	+++ +	+++ +
Flavobacterium okeanokoites	+++ +	++++ +	+++ +	++++ +	++++ +	+++ +	++	+++ +	+++ +	−	−	+++ +
Flavobacterium uliginosum	+++ +	++++ +	+++ +	++++ +	++++ +	+++ +	−	+++ +	+	−	−	−
Micrococcus infimus	+	+++ +	A	−	+		−	−				
Photobacterium splendidum	+++ +	+++ +	A	++	+++		−	+				
Pseudomonas pleomorpha	++	++++ +	A	++	+++		−	++				
Pseudomonas vadosa	++	++++ +	++++ +	++	++++ +	+++ +	−	+++ +	+++ +	−	−	+++ +

Table 2 (continued)

RELATIVE TURBIDITY CAUSED BY THE MULTIPLICATION OF MARINE BACTERIA IN NUTRIENT BROTH FOR 6 DAYS AT 20°C, 4 DAYS AT 30°C, OR 1 DAY AT 40°C AT DIFFERENT HYDROSTATIC PRESSURES

Culture	300 Atm			400 Atm			500 Atm			600 Atm		
	20°C	30°C	40°C	20°C	30°C	40°C	20°C	30°C	40°C	20°C	30°C	40°C
Pseudomonas xanthochrus	+++	+++	A	+++	+++	+	++	++	++	+	+	−
Vibrio hyphalus	+	+	A	+	−	−	−	−	−	−	−	−
Mixed microflora from mud	+++	+++	+++	+++	+++	+++	+++	+++	+++	+++	+++	+++
	+	+	+	+	+	+	+	+	+	+	+	+

Note: All cultures listed above, except those marked "A" which failed to grow at 40°C, showed four plus (+ + + +) growth in the controls incubated at normal pressure.

From ZoBell, C. E. and Johnson, F. H., *J. Bacteriol.*, 57, 179, 1949. With permission.

Table 3

MPN OF DIFFERENT PHYSIOLOGICAL TYPES OF
BACTERIA DETECTED PER GRAM OF WET SEDIMENT
FROM THE PHILIPPINE TRENCH IN SELECTIVE MEDIA
INCUBATED AT DIFFERENT HYDROSTATIC
PRESSURES AT 3 TO 5°C

	Galathea Station							
	No. 418		No. 419		No. 420		No. 424	
Latitude	10°13′ N		10°19′ N		10°24′ N		10°28′ N	
Longitude	126°43′ E		126°39′ E		126°40′ E		126°39′ E	
Water depth	10,190 m		10,210 m		10,160 m		10,120 m	
	Incub. pressure (atm)							
	1	1,000	1	1,000	1	1,000	1	1,000
Total aerobes	10^3	10^6	10^3	10^5	10^4	10^5	10^1	10^6
Total anaerobes	10^3	10^5	10^4	10^5	10^3	10^5	10^4	10^5
Starch hydrolyzers	10^2	10^3	10	10^2	10	10^2	10^2	10^3
Nitrate reducers	10^2	10^5	10^2	10^4	10^3	10^5	10^2	10^5
Ammonifiers	10^2	10^5	10^3	10^4	10^3	10^5	10^3	10^5
Sulfate reducers	0	10^2	0	10	0	10^2	0	0

From ZoBell, C. E. and Morita, R. Y., *J. Bacteriol.*, 73, 563, 1957. With permission.

Table 4

SINKING RATES OF DEAD MARINE
PROTOZOANS, FISH EGGS, ZOOPLANKTERS,
AND LIVING AND DEAD PHYTOPLANKTERS

Group	Rate (m/day)	No. of species used
Fish eggs	215—400	2
Phytoplankton		
1. Living	0—30	~25
2. Palmelloid stages	~5—6,150	1
3. Dead	<1—510	~10
Protozoans		
1. Foraminifera	30—4,800	—
2. Radiolarians	~350	—
Zooplankton		
1. Amphipoda	~875	1
2. Chaetognatha	~435	1
3. Cladocera	~120—160	2
4. Copepoda	36—720	14
5. Heteropoda	~1,400	1
6. Ostracoda	400	1
7. Pteropoda	760—2,270	5
8. Salpa	165—253	2
9. Siphonophora	240	1
10. "Animal plankton"	~225—500	?

From Smayda, T. J., *Limnol. Oceanogr.*, 14, 621, 1969. With permission.

Table 5
THRESHOLD CONCENTRATIONS OF THREE
GROWTH-LIMITING SUBSTRATES (IN mg/l) IN
SEAWATER AT SEVERAL RELATIVE GROWTH
RATES OF SIX STRAINS OF MARINE BACTERIA
AND THE CORRESPONDING MAXIMUM
GROWTH RATES (IN HR^{-1})

Strain	D/μ_m	Lactate	Glycerol	Glucose
208	0.5	0.5	1.0	0.5
	0.1	0.5	1.0	0.5
	0.05	1.0	5.0	1.0
μ_m		0.15	0.20	0.34
102	0.3	0.5	no growth	0.5
	0.1	1.0		5.0
	0.05	1.0		10.0
μ_m		0.45		0.25
101	0.5	5	5	
	0.3	10	10	no growth
	0.2	20	100	
	0.1	50		
μ_m		0.45	0.60	
204	0.3	1	5	1
	0.1	5	5	5
	0.05	10	10	20
μ_m		0.15	0.35	0.40
317	0.5	20	50	
	0.3	50	50	no data
	0.1	100	>100	
	0.05	>100		
μ_m		0.85	0.70	
201	0.5	20	50	20
	0.3	50	50	50
	0.2	100	>100	>100
	0.1	>100		
μ_m		0.80	0.65	0.80

Note: Strain 208: *Achromobacter aquamarinus*
Strain 102: *Spirillum lunatum*
Strain 101: *Spirillum* sp.
Strain 204: *Vibrio* sp.
Strain 317: *Achromobacter* sp.
Strain 201: *Pseudomonas* sp.

From Jannasch, H. W., *Limnol. Oceanogr.*, 12, 264, 1967. With permission.

Table 6
OXIDATION OF REDUCED SULFUR
COMPOUNDS AND FIXATION OF $^{14}CO_2$ BY
MARINE STRAIN D10-B

Substrate	O_2 absorbed, μmol/120 min	$^{14}CO_2$ fixed, μmol/120 min
$S_2O_3^{2-}$	7.50	0.67
S^0	5.27	0.43
SO_3^{2-}	3.97	0.07
$S_4O_6^{2-}$	3.89	0.29
None	1.15	0.10

Note: Each Warburg flask contained potassium phosphate buffer, pH 7.5, 150 μmol; NaCl, 1.55 mmol; cells, 4.25 mg protein; NaH$^{14}CO_3$, 2.27 μmol (1.15 × 10^5 counts/ minute per μmol); one of the following substrates: S^0 (wetted with Tween-80®, 0.005%), 10 mg; SO_3^{2-}, 10 μmol; $S_2O_3^{2-}$, 10 μmol; $S_4O_6^{2-}$, 10 μmol; and water to make a final reaction volume of 3.0 mℓ.

From Adair, F. W. and Gundersen, K., *Can. J. Microbiol.*, 99, 1, 1974. With permission.

Table 7
COMPARISON OF THE HETEROTROPHIC
GROWTH OF *T. NOVELLUS* ATCC **AND**
THIOBACILLUS A2

Substrate	T. novellus ATCC	Thiobacillus A2
D-Glucose	+	+
D-Fructose	+	+
Sucrose	−	+
D-Gluconate	+	+
D-Mannitol	+	+
Lactose	−	−
Maltose	+	+
D-Ribose	Weak	Weak
Glycerol	+	+
Pyruvate	+	+
Formate	+	+
Acetate	+	+
Propionate	Weak	+
Butyrate	−	+
Methanol	+	+
Ethyl alcohol	+	+
n-Propanol	+	+
n-Butyl alcohol	−	+
L-Histidine	+	+
Proline	+	+
L-Leucine	−	+
L-Isoleucine	−	+
L-Tryptophian	−	−
DL-Serine	−	−
L-Alanine	Weak	+
L-Glutamate	−	+
L-Aspartate	−	+
Malate	−	+
Succinate	−	+
Citrate	−	−
Benzoate	−	+
p-Hydroxybenzoate	−	+
m-Hydroxybenzoate	−	+
p-Aminobenzoate	−	−
Cyclohexanol	−	−
Cyclohexane carboxylate	−	+

From Taylor, B. F. and Hoare, D. S., *J. Bacteriol.*, 100, 487, 1969. With permission.

Chapter 3*

DYNAMICS OF ORGANIC MATERIALS AND MICROORGANISMS

I. DECOMPOSITION OF PARTICULATE ORGANIC MATERIALS

All types of living organisms die at the same rate as they are born, whereby a continuous return of the raw materials is possible by biological transformation. In this process, bacteria and allied microorganisms play many beneficial roles in nature. They are responsible for most of the decomposition of dead animal and plant bodies, thus returning important nutrients to the primary producers. Animals as well as higher plants may take part in the general cycle but they are not indispensable transformers. A reducing cycle cannot be completely performed without the aid of bacteria. Indeed, considerable quantities of plant materials must, at times, bypass animals and be decomposed directly to inorganic compounds by bacterial activities.

Many microorganisms in aquatic environments are extremely versatile in their ability to decompose different organic compounds, and any organic compound that occurs naturally can be oxidized by some type of microorganism. Recently it has been shown that even pesticides and plastics, unknown in the biosphere before this century, are attacked slowly by microorganisms in soil and aquatic environments. The enormous catalytic power of bacteria contributes to the major chemical transformations occurring on the surface of the Earth; this is because of their rapid rate of reproduction in favorable environments and their high metabolic potentials, i.e., on a unit body weight basis, the respiratory rates of some aerobic bacteria are determined to be hundreds of times greater than that of man.

A fraction of the organic debris produced by the food chain tends to accumulate to a steady-state value several orders of magnitude greater than the organisms from which it is derived. Essentially there are two natural processes which contribute to the formation of these organic products in aquatic environments.[32]

The first of these processes involves the decomposition of plant and animal materials. This may result in an initial 15 to 50% loss of total biomass due to post-mortem changes in the permeability of cell membranes and the effects of autolytic enzymes. Products from dead organisms include amino acids, keto acids, and fatty acids. The more refractory components of organisms, such as cellulose and chitin, may be decomposed almost exclusively by bacterial action.

The second process involved in the formation of organic materials is the liberation of extracellular products by living plants and animals. From 5 to 30% of the extracellular products of photosynthetically fixed carbon dioxide may be released as soluble organic carbon. Some of the common phytoplankton species, as well as natural populations in general, release extracellular products from healthy cells amounting to between 15 and 35%. Substances which have been identified as resulting from the release of extracellular products include polysaccharides, polypeptides, amino acids, and glycolic acid, as well as certain species-specific biologically active compounds such as enzymes. The release of amino acids by healthy zooplankton also has been reported in quantities which, within 1 month, would be sufficient to replace the quantities found to occur naturally in oceanic waters. Thus the release of organic compounds to the carbon reservoir in an aquatic environment, either through decomposition processes or from healthy plankton, could be quite appreciable. For an average annual marine primary production of 70 g $C/m^2/year$ this would amount to an annual minimum input of 18 to 42 g $C/m^2/year$.

---

* All figures and tables for Chapter 3 appear after the text.

The bodies of living animals and plant cells, of course, provide favorable environments for the activity and growth of heterotrophic microorganisms. They supply desirable organic nutrients and relatively constant environmental conditions for these microorganisms (Figure 1). Bacteria have been shown to live inside cells of higher
organisms, sometimes even inside the nucleus (Figure 2), and probably in a commensal
relationship. Under these conditions nutrients cannot become limiting factors for the
survival and multiplication of these microorganisms. After the animal and plant cells
die there is an initial period of autolysis when soluble materials do not leach out.
During this stage a certain number of bacteria may invade through the injured part of
the cell wall of the dead cell to participate in accelerating the decomposition (Figure
3). At the next stage of decomposition, soluble materials leach out from the dead cell,
whereby a number of bacteria in the matrix of dead protoplasm form an aggregate in
water environments (Figure 4). Colonized microorganisms on the surface of cell walls
or other insoluble dead materials begin to render these insoluble materials soluble by
enzyme action. Then populations of predators such as ciliates and nematodes begin to
build up, especially in the case of degradation of multicellular organisms (Figure
5).[141,143] In the particular case of dead eucaryotic plants, the plants have cell walls that
confer shape and rigidity to their cells. These walls are usually much thicker than those
of procaryotic cells and are composed of cellulose. Cellulose is a polymer consisting
entirely of glucose units connected in a β-1,4 linkage:

A single cellulose polymer may have up to 15,000 glucose units, corresponding to a
length of several μm; thus its molecular weight ranges from 200,000 up to 2,400,000.
These macromolecules intertwine to produce microfibrils 5 to 50 nm in diameter which
form the structure of the cell wall. A number of other polysaccharides also are associated with the cellulose of the plant cell wall. Cellulose also is found in the cell walls of
certain lower fungi; although in some other lower fungi, or in higher fungi, the basic
cell wall structure is composed of a β-1,3 linkage. Among protozoans, dinoflagellates
and a few ciliate species are known to have cellulose cell walls.

Among glucose polymers, cellulose is relatively more resistant to enzymatic and microbial breakdown and is thereby less rapidly hydrolyzed. All microorganisms that can
utilize substances of high molecular weight excrete enzymes onto the substrate, and
these enzymes break the substrate down into small assimilable molecules since molecules larger than those with 100,000 mol wt would not be able to penetrate the cell
membrane. Hence, in cellulolytic microorganisms, the catalytic enzymes that convert
cellulose to each glucose unit are of three types: C_1 cellulase, C_x cellulase, and β-glu-
cosidase.

C_1 cellulase is a collective name for either surface-located or extracellular enzymes
found in true cellulolytic microorganisms and it acts only on native cellulose. These
enzymes, which are capable of cleaving high molecular substrates, must get out of the
cell into the environment or reside on the cell surface, for the substrate cannot per-
meate into the cell due to its high molecular weight. When C_1 cellulase is an extracellular enzyme, in order to be sufficiently effective it must remain in the excreted place
without diffusing away from the cellulolytic microorganisms. C_1 cellulase has little or
no action on partially degraded cellulose, i.e., on intermediates in the conversion of

the polymer (cellulose) to the monomer (glucose). A single population commonly excretes more than one C_1 cellulase; these C_1 enzymes have slightly different structures, but all presumably function in the same way.

C_x cellulase cannot hydrolyze native cellulose but instead cleaves the partially degraded polymers. C_x cellulase also is a type of either surface-located or extracellular enzyme, and it is more widespread among various species of bacteria, actinomycetes, and fungi than among truly cellulolytic microorganisms. C_x cellulase is a collective name for β-1,4,glucanase, and two types are distinguished: one type of catalyst, known as endoenzyme, acts on the linkages between glucose units within the long polysaccharide chain; the second type of catalyst, known as exoenzyme, acts on the linkages only at the ends of chains. The hydrolysis products of endoenzyme are cellobiose, various oligomers, and sometimes glucose. The hydrolysis product of exoenzyme is usually cellobiose.

The last phase in the breakdown of cellulose to glucose is catalyzed by β-glucosidase. This hydrolyzes all intermediates in the breakdown of cellulose, such as cellobiose, cellotriose, and other oligomers with low molecular weight.

Cellulose is insoluble in water and forms long fibrils. Microorganisms which hydrolyze cellulose are found closely associated with cellulose, and they must excrete extracellular or surface-located enzymes in order to make the substrate available by converting the cellulose from insoluble material to soluble saccharides that can penetrate the cell membrane.

Cellulase is inducible in most microorganisms and is synthesized in the presence of cellulose or carbohydrates that are structurally similar to this polysaccharide. Cellulolytic microorganisms are widely distributed in nature. Among the cellulolytic species are aerobic and anaerobic bacteria, actinomycetes, filamentous fungi, and basidiomycetes. A diverse group of fungi has been shown to utilize cellulose for its carbon and energy sources in the terrestrial region and to be the main agents of cellulose degradation in humid soils. The cellulolytic microorganisms in aquatic environments are restricted to only a few groups of bacteria and possibly some actinomycetes. These bacteria of greater significance in cellulose degradation are *Bacillus*, *Cellulomonas*, *Clostridium*, *Corynebacterium*, *Cytophaga*, *Polyangium*, *Pseudomonas*, *Sporocytophaga*, and *Vibrio*. Aerobic bacteria generally only convert cellulose to major products of carbon dioxide and cell substance, and they rarely excrete organic acids and other intermediates at appreciable levels into their environment. This is because the initial hydrolysis of cellulose is the rate-limiting reaction in its degradation. Anaerobic fermentation of cellulose is carried out by many *Clostridium* species that commonly inhabit lake sediments, river mud, sewage, and marine sediments. The main products excreted into their anaerobic environment are carbon dioxide, hydrogen gas, ethanol, acetic acid, formic acid, succinic acid, butyric acid, and lactic acid.

Fluctuation of the cellulose stock in the hydrosphere may be most highly dynamic in zones of littoral vegetation. For example, in a particular water chestnut ecosystem (area 2 km²; depth 0.5 to 1.5 m) in a hypereutrophic lake, Lake Kasumigaura, Japan, the cellulose stock in the live vegetation is sedimented onto the lake bottom after death of the plants, forming the equivalent cellulose stock in plant residues. During the sedimentation, rapidly decaying compounds should have been released into the ambient water or converted into bacterial cell substance.[142] The major stock of cellulose seems to be kept in the surface sediment and to be in a steady-state equilibrium with a time scale of 1 year (Figure 6). This equilibrium results from the input of cellulose products of the water chestnuts and eucaryotic phytoplankters in the water column and from the output of the cellulose decomposition by cellulolytic microorganisms. However, the increase of cellulose concentration apparent in the sediment layer depends on the historical progress of eutrophication in the overlying water. Therefore, such littoral vegetation in hypereutrophic waters should be active and efficient in irreversibly removing nutrients from the hydrosphere (water column) to the lithosphere (bottom sediment). This process of removing nutrients works as a negative feedback system in the steady-state equilibrium of nutrient concentration in the lake for the following reason. The littoral vegetation grows extensively in eutrophic lakes and produces organic compounds which are relatively resistant to microbial decomposition; thus the biological elements stored in these plants are removed from the cycling of biological elements in the hydrosphere. Most eucaryotic phytoplankters in oliogotrophic or mesotrophic water masses, on the other hand, might take up nutrients from the hydrosphere, but these are decomposed into inorganic nutrients very efficiently as they are mainly composed of rapidly decaying compounds. Cellulose and other slowly decaying compounds in the plant residues on the surface sediment form the primary energy source of the detritus food chain in the ecosystem, as they are attacked by bacteria and allied microorganisms. These microorganisms are consumed in turn by predatory benthic animals such as nematodes and benthic shrimp,[143] and thereby the biomass contained in the ecological pyramid is inverted due to the very low primary productivity in winter. Thus, the dead water chestnuts have been shown to contribute annually to organic sedimentation of cellulose and other slowly decaying compounds on the lake bottom.

At the end of June when the water chestnuts reach their maximum standing crop of the year, blue-green algae belonging to the genera *Microcystis* and *Anabaena* form an algal bloom in the ecosystem. Eventually, these blue-green phytoplankters completely cover the leaves of the water chestnuts, causing their death (Figure 7), and their dead fragments start to sediment onto the lake floor at the end of September. During the period from the formation of the blue-green algal bloom to the sedimentation of the dead water chestnuts, eucaryotes become only very minor components in the phytoplankton flora, and therefore cellulose produced by or existing in the cells of these eucaryotic phytoplankters is negligible in the water column. Largely because horizontal movement of the lake water is made very sluggish by the barrier effect of the water chestnuts (several months being required for water renewal), the *in situ* cellulolytic processes in the lake during this period can be analyzed in detail by applying a mathematical model for cellulose dynamics in hypereutrophic lakes.

Generally, the cellulolytic activity is governed by a number of environmental influences. The major environmental factors are believed to be the available nitrogen level, temperature, aeration, water activity, pH, Eh, and the presence of other chemical compounds in the plant residue. Even when these environmental factors are simulated in laboratory experiments some other unknown natural factors might affect the cellulolytic process to produce great differences between *in situ* and laboratory results. Accordingly, the best approach is to study the dynamics *in situ* when environmental conditions are satisfactory.

The standing stock of cellulose in water of the water chestnut ecosystem started to increase during early summer through the active formation of detrital cellulose from the water chestnuts and eucaryotic phytoplankton. During late summer and autumn the cellulose stock decreased due to active microbial decomposition; at this time little detrital cellulose was formed by the water chestnuts and eucaryotic phytoplankton because of the predominance of *Microcystis* in the phytoplankton community (Figure 8). Cellulolytic microorganisms associated with detrital particles in lake water can be differentiated into two types: Type 1 which can utilize cellulose as the sole energy source; and Type 2 which cannot utilize cellulose as the sole energy source but which excretes cellulase. In order to enumerate the Type 1 and Type 2 microorganisms, agar plate media were used consisting of 1.5% cellulose mikrokristallin (Art. 2330, Merck®) and 1.5% cellulose mikrokristallin (Art. 2330, Merck®) with 0.01% yeast extract, respectively. Almost all the cellulose particles in the water column were microscopic, measuring between 0.45 and 40 μm. The standing stock of cellulose included those cellulose particles which had settled down immediately above the lake bottom; the settled cellulose amounted to less than 5% of that in the whole water column throughout the study period (Figure 6). The dynamics can be expressed by using the following equations.

The nature of cellulose fluctuation is approximated by the following expression with a correlation coefficient of $r = 0.97$:

$$\log y = 3.004 + 0.706 \sin \frac{\pi}{47.0} (t - 75.0)$$

where y is the concentration of cellulose (mg C/m^2), and t is days since May 1, 1973. The fluctuation of cellulose leads to density fluctuations of cellulolytic bacteria, firstly for Type 2 by the following equation, with a correlation coefficient of $r = 0.99$ and a time lag of 5.1 days, as

$$x' = 8.772 + 6.975 \sin \frac{\pi}{67.0} (t - 87.5)$$

where x' is the density of cellulolytic bacteria Type 2 ($\times 10^7/m^2$); and secondly for Type 1 by the following equation with a correlation coefficient of $r = 0.87$ and a time lag of 16.1 days, as

$$x = 8.749 + 5.496 \sin \frac{\pi}{61.5} (t - 98.5)$$

where x is the density of cellulolytic bacteria Type 1 ($\times 10^7/m^2$). Representatives of these cellulolytic bacteria, both Type 1 and Type 2, belonged to the genera *Cytophaga*, *Vibrio*, and *Pseudomonas*. The population density of cellulolytic fungi was less than $1/\ell$ of lake water during the study period.

These relationships show that an increase in dead eucaryotic plants leads to an increase in cellulolytic bacteria of Type 2 with a time lag of 5.1 days. Thereafter, easily metabolizable compounds were assimilated by heterotrophs in 11.0 days, and cellulose still left after this time was attacked mainly by cellulolytic bacteria of Type 1.

Cellulolytic rate (dy/dt) can be defined by a differential equation based on the standing stock of cellulose:

$$\frac{dy}{dt} = \frac{0.706}{47.0} \cos \frac{\pi}{47.0} (t - 75.0) \times 10^{\left[3.004 + 0.706 \sin \frac{\pi}{47.0} (t - 75.0) \right]}$$

Biochemical processes are influenced profoundly by temperature, among other environmental factors. Thus cellulolytic rate at 20°C (dy$_{20}$/dt), as corrected by the Arrhenius equation, was used to obtain the theoretical relationship between the cellulolytic bacteria and their activity:

$$\frac{dy_{20}}{dt} = \frac{e^{-E/293.2R}}{e^{-E/RT}} \times \frac{dy_T}{dt}$$

where T is absolute temperature, R is the gas constant, and E is the constant related to the energy of activation of the molecules in the system. The best regressions between dy$_{20}$/dt and cellulolytic bacteria were obtained firstly for Type 2 (X') with a time lag (B$_2$) of 8.2 days

$$\frac{dy_{20}}{dt} = 2.9970 \, X' - 26.505$$

giving the best correlation coefficient of $r_{2\ max} = 0.9003$, with $E_2 = 22,401$; and secondly for Type 1 (X) with a time lag (B_1) of 16.4 days,

$$\frac{dy_{20}}{dt} = 3.3709\ X - 27.194$$

with the best correlation coefficient of $r_{1\ max} = 0.9047$, with $E_1 = 22,401$. These results indicate that the cellulolytic activity was initiated by Type 2 when the bacterial density had attained a threshold density in the lake water of $8.87 \times 10^7/m^2$ at $dy_{20}/dt = 0$. This bacterial density, estimated by the agar plate method, actually indicates the number of particles on which one or more bacteria were adsorbed and could multiply in the medium.[144] The time lags, B_1 and B_2, could be attributed to the time required thereafter for bacterial multiplication to form a larger bacterial colony on each cellulose particle. Laboratory experiments show that most cellulolytic bacteria isolated from the lake can multiply from 1 cell to 10^8 cells during 20 days of incubation with cellulose as the sole nutrient. These Type 2 bacteria multiply 1.97×10^3 times in 8.2 days (B_2). The same approach can be applied to cellulolytic bacteria of Type 1; they could multiply 3.82×10^6 times in 16.4 days (B_1). Thus the calculated cellulolytic rate corresponds to the bacterial number in each bacterial particle (or colony), consisting of 1.97×10^3 cells for Type 2 or 3.82×10^6 cells for Type 1. Therefore, cellulolytic rate can be estimated as follows:

1. Cellulolytic rate by Type 2 = 2.997×10^{-7} mg C/cellulolytic bacterial particle per day = 5.910×10^{-10} mg C/cellulolytic bacterial cell per day
2. Cellulolytic rate by Type 1 = 3.371×10^{-7} mg C/cellulolytic bacterial particle per day = 1.288×10^{-12} mg C/cellulolytic bacterial cell per day

This analysis on cellulolytic activity *in situ* in the hypereutrophic lake shows that the turnover time of cellulose should be approximately 20 days at the optimal conditions for cellulolytic bacteria of Type 2. In the natural condition, however, easily metabolizable organic materials may be completely utilized within several days after the eucaryotic cells die. Accordingly, the cellulolytic flora must show succession from Type 2 to Type 1; thereby the cellulose decomposition can be completed within 200 days.

Just as cellulose is a moderately refractory component of eucaryotic plants, chitin has been shown to be another moderately refractory component of animals. Chitin is the constructural constituent giving mechanical strength to the rigid exoskeletons of insects, crustaceans, some mollusks, coelenterates, protozoans, and certain other groups of animals. The cell walls of many fungi also contain chitin. Chitin is found only in a few species of protozoans, but a protein-polysaccharide matrix (pseudochitin) is present in some species.

Chitin is a crystalline compound which is insoluble in water. Chitin consists of a long chain of a polymer of *N*-acetylglucosamine molecules in β-1,4,linkage. The polymer is similar to cellulose, but one of the hydroxyl groups of each glucose in cellulose is replaced by an acetylamino unit:

Chitin is distributed widely in aquatic environments, where major stocks of chitin can be attributed to the exoskeletons of planktonic crustaceans; i.e., this substance forms the major part of the organic matrix on which calcium carbonate is deposited to form the exoskeleton. Copepods, a predominant group of marine zooplankton, alone have been calculated to produce billions of tons of chitin every year.

The catalytic enzymes of chitinoclastic microorganisms include two types, chitinase and chitobiase, which convert chitin to N-acetylglucosamine units. Chitinase is an extracellular enzyme and usually inducible. Chitinase is a collective name for endoenzymes that catalyze a depolymerization of the chitin chain to yield chitobiose and oligomers. Oligomers have several units of N-acetylglucosamine. Chitobiase hydrolyzes the oligomers and chitobiose to yield N-glucosamine. Hence chitobiase also is called N-acetylglucosaminidase, a name which may be more appropriate since this enzyme catalyzes not only chitobiose but also oligomers to yield N-acetylglucosamine. The N-acetylglucosamine then is converted to acetic acid and glucosamine. The glucosamine and acetic acid are readily utilized as sources of materials and energy by a number of organisms. Thus the biochemical breakdown of chitin involves the conversion of an insoluble molecule into water soluble products that can penetrate microbial cells and serve to provide energy and material.

Chitinoclastic microorganisms are distributed widely in nature. They are bacteria, actinomycetes, and fungi. In most terrestrial regions the important chitinoclastic microorganisms have been shown to be actinomycetes. Among bacteria, chitin utilization is attributed to bacterial species mostly in the genera *Beneckea, Vibrio, Pseudomonas, Cytophaga, Flavobacterium, Micrococcus, Aeromonas, Achromobacter,* and *Clostridium. Clostridium* can metabolize the chitin molecule in the absence of oxygen gas.

Fluctuation of the chitin stock may be highly dynamic in eutrophic regions. Arthropods are the predominant secondary producers in the water chestnut ecosystem of Takahama-iri Bay, Lake Kasumigaura, and are partly macrozooplankters such as freshwater shrimps, *Macrobrachium nipponense* and *Neomysis intermedia*; partly larvae of midges, *Chironomus plumosus* and *Tokunagayusurika akamusi*; and partly microzooplankters, *Diaphanosoma brachyurum, Scaphaleberis macronata, Ceriodaphnia rigaudi, Moina rectirostris, M. dubia, Bosmina coregoni, B. fatalis, Bosminopsis deitersi, Chydrus sphaericus, Cyclops vicinus, Mesocyclops leuckarti,* and *Thermocyclops taihokuensis*. The production of chitin by these arthropods is high. Daily growth rates of the microzooplankters, *Macrobrachium nipponense*, and *Neomysis intermedia* have been determined to be approximately 33%, 2%, and 12%, respectively. Because of the high standing stock of chitin in arthropod biomass, high production rates of chitin and unstable population dynamics, a large amount of chitin in the form of carcasses and molts enters into a detrital chitin stock in the ecosystem (Figure 9). A large amount of chitin is produced intensively in late spring and summer and is decomposed actively in summer. From autumn to next spring, the input and output of chitin in the ecosystem decreased, probably due to lower water temperature resulting in the small amplitude of steady-state oscillation in the standing stock of chitin. The annual budget of chitin shows a surplus of input into the ecosystem resulting in the accumulation of large amounts of chitin in the sediment. This process in an eutrophic environment is in contrast to oligotrophic environments where little chitin accumulates in the sediment.[145] Therefore, the secondary producers make a considerable contribution to the life process of the ecosystem through organic sedimentation of a refractory material like chitin in just the same way as primary producers contribute to cellulose accumulation in the lake sediment

The density of chitinoclastic bacteria was related to the concentration of particulate chitin in the water chestnut ecosystem (Figure 10). During the summer period of oxygen deficiency when chitin input into the ecosystem was negligible, an increase of detrital chitin led to an increase of chitinoclastic bacteria Type 2 with a time lag of approximately 5 days, and then with a time lag of approximately 20 days, to chitinoclastic bacteria Type 1. Chitin in fresh carcasses, together with many other chemical components, started to be decomposed by chitinase excreted mainly by chitinoclastic bacteria Type 2. After almost all the easily metabolizable components of the corpses were assimilated by heterotrophs within 20 days, any chitin still remaining was attacked mainly by chitinoclastic bacteria Type 1. During this period, daily chitinoclastic rates can be estimated using a mathematical model.[143] The turnover rate of chitin *in situ* (Table 1) is shown to be similar to that experimentally measured in batch cultures of purified chitinoclastic bacteria grown in a chitin medium prepared with lake water. This is probably because the microenvironment on the surface of chitin is modified into a favorable environment for chitinoclastic bacteria which is little affected by the difference of ambient waters; i.e., it seems that the artificial water in the laboratory as compared with the natural water does not appreciably affect microbial activity in the microenvironment. Daily chitinolytic rate per bacterial count was highest during the summer at 30°C, when the maximum rate was 1.3×10^{-7} mg/bacterial count/day. Similar very active biodegradation of chitin has been measured in a marine eutrophic region; *in situ* rates of native chitin degradation in a salt marsh environment have been determined to be as high as 87 mg/day/g chitin.[146] These rates are 20 times greater than those measured in coastal regions of the Pacific Ocean.[147,148] The chitinoclastic rates during other seasons in the water chestnut ecosystem of Takahama-iri Bay fluctuated little, remaining around 5.3×10^{-9} mg/bacterial count/day. This rate is almost the same as that measured in coastal regions of the Pacific Ocean,[147] where the rate is 3.0×10^{-9} mg/bacterial count/day. Chitinoclastic activity at similar rates of 1.9×10^{-9} to 3.1×10^{-9} mg/bacterial count/day at 20°C was measured for *Vibrio* species from the Puget Sound estuary, another coastal area of the Pacific Ocean.[149] These rates are equivalent to between 1.0 to 3.0 mg/day/g chitin, assuming that the average chitin particle in aquatic environments measures less than 0.033 cm. Chitin decomposition in oceanic regions has been shown to be performed by most of the heterotrophic bacteria, i.e., chitinase is inducible by most marine heterotrophic bacteria.[147] These chitinoclastic bacteria are found in any water mass of the surface layer, the intermediate layer, and the deep layer of the Pacific Ocean.[147] Some of them may be in the form of bacterioplankton, but those participating in chitin decomposition should be on chitin particles. A considerable number of chitinoclastic bacteria already are attached to living plankton in natural waters. They number from 10^6 to 10^7/g (dry weight) of zooplankton. Therefore, a considerable number of chitinoclastic bacteria already are attached to molted exoskeletons and to dead planktonic crustaceans. Since dead plankton accumulate mostly at the boundary surface of different water masses, their decomposition in these boundary layers would be complete within about 140 days, with none accumulating in deep sediments. In the oceanic region, in general, chitin may be completely mineralized within 140 days in the surface layer at 15°C, within 370 days in the intermediate layer at 5°C, and within 500 days in the deep waters at a few degrees centigrade.[147] Morita[51] has isolated from various locations of the hydrosphere, including the Antarctic Ocean, many psychrophilic chitinoclastic bacteria which can actively digest chitin at temperatures slightly above and below 0°C. This kind of study lends support to the belief that active chitinolytic activity can occur in any part of the aquatic environment.

Although lipids exist mostly in liquid phase in natural waters, the solubility of lipids is very low and lipid droplets are suspended in water or form surface films on water in aquatic environments. The lipolytic microorganisms in association with the droplets are concentrated only at the lipid-water interface and not within the droplets.[150] Accordingly, the process of lipid decomposition may be the same in principle as that of decomposition of cellulose of chitin particles.

Lipids are estimated to be contained in marine plankton in amounts of 2 to 20%. The concentration of lipids in seawater ranges between 0.01 to 0.12 mg/ℓ for Pacific Ocean water, which indicates that enormous amounts of lipids are contained in the marine environment.[15] Natural slicks, smooth glassy streaks, or patches on the sea surface are contaminated films of organic oil probably derived primarily from diatoms which contain droplets of oil in their cells to assist flotation and as an emergency food supply.

Fatty acids found in lipids of marine fishes are all of the common straight chain and monocarboxylic series. They contain only even numbers of carbon atoms, varying from 12 to 26 and sometimes to 28. Marine fish lipids can also be found in marine phytoplankters and planktonic crustaceans, and as these small herbivorous filter feeders are the main food link between aquatic plants and animals, the typical structures of marine triglycerides originate in phytoplankton and are to a large degree retained through the marine food chains.[152] This transfer of lipids has also been shown experimentally, although a slight modification in the composition of fatty acids is apparent for lipids of each trophic level — diatom (*Chaetoceros*) → crustacean (*Artemia*) → fish (*Lebistes*) — in the food chain (Table 2).[153]

Lipids produced by plankton, nekton, and benthos are decomposed finally by lipolytic microorganisms. Lipase is released as an exoenzyme from the microorganisms and hydrolyzes glycerides resulting in the production of glycerol and free fatty acids. Lipase hydrolyzes most triglycerides, but is less active against diglycerides and does not hydrolyze monoglycerides at an appreciable rate. During the hydrolysis, an exchange takes place between liberated fatty acids and the glyceride fatty acids. Lipase also is known as a very stable enzyme and is resistant to most enzyme inhibitors. Most lipolytic mocroorganisms utilize the glycerol resulting from the hydrolysis of lipids as a source of energy and materials.[154] The liberated fatty acids are soluble in water and are utilized by most microorganisms. These fatty acids are oxidized biochemically by a process called beta oxidation. Complete oxidation of a fatty acid yields high energy, resulting in the synthesis of many ATP molecules.

Important lipolytic microorganisms in the aquatic environment include the genera *Pseudomonas, Vibrio, Seratia, Achromobacter, Bacillus,* and *Sarcina.* The utilization of lipids by anaerobic microorganisms in sediments of aquatic environments is known, especially in relation to the formation of petroleum.

Some parts of lipids may be trapped with dead organisms in sediment. Most of the lipids, however, may be released as an emulsion from dead organisms still suspended in the water column. These lipids in emulsion may be exposed to hydrolysis by lipolytic microorganisms while floating up in the water column, and part of them may reach the surface and form a natural surface film. Such films, known as slicks, may hinder the gaseous exchange between the atmosphere and the hydrosphere, especially in nursery ponds or after the phytoplankton bloom in natural waters.

Lipid decomposition experiments with marine bacteria show that lipolytic activity may be almost of the same rate whether the bacterial contact with the lipid emulsion occurs on particulate matter, on plankton, in seawater, in sediments, or in the surface film. From *in situ* observation of the distribution of lipolytic microorganisms and from a series of experiments with isolated marine lipolytic bacteria, we know that lipids contained in particulate matter are gradually decomposed in accordance with the biodegradation of other organic materials in the particle. The complete decomposition of lipids in a particle takes place within 40 days at the optimal conditions for lipolytic microorganisms in the marine environment. A fraction of lipids in particulate matter may be released as small emulsions; these are estimated to be decomposed within 2 days by lipolytic microorganisms which also are released at the same time from the particulate matter.[150]

II. ASSIMILATION AND DECOMPOSITION OF ORGANIC SOLUTES BY MICROORGANISMS

The concentrations of dissolved organic compounds in aquatic environments are many times lower than concentrations inside the cells of most microorganisms, therefore concentrations in natural waters have reached some threshold level that permits the support of only microorganisms which have efficient uptake systems of these compounds. Three categories of heterotrophic bacteria are involved in the major utilization of these organic solutes in aquatic environments:

1.   Free-living bacterioplankton with active transport systems for highly efficient uptake of organic compounds
2.   Stalked bacteria adherent to the interfaces between the liquid phase and solid or gaseous phases
3.   Bacteria tending to clump or form aggregates that include organic materials

From direct microscopic examination of freshwater or marine materials, it has been estimated that approximately 95% of aquatic bacteria are Gram-negative.[155,156] The area between the outer lipopolysaccharide layer and the plasma membrane of Gram-negative bacteria is called the periplasmic space and contains a number of specific proteins involved in the transport of substrates. These proteins are called binding proteins because they specifically bind the substrates that they transport. Binding proteins can leave the cell readily, and they are believed to function in the initial stages of transport by binding the substrate and bringing it to the membrane carrier which is firmly embedded in the membrane. There the substrate combines with a membrane-bound carrier protein which then releases the substance unchanged inside the cell. This mechanism of substrate transport implies a process more complex than simple diffusion. Hence it is called facilitated diffusion; it is distinguished from simple diffusion by evidence of saturation kinetics and stereospecificity, indicating interaction of the substrate with a component of the membrane. The rate of facilitated diffusion is greater than would be expected for passive diffusion across a lipid phase. However, by definition the process involves no energy input and results only in equilibration of the substrate across the membrane in accordance with the electrochemical potential.[157] When energy-generating systems couple with the facilitated diffusion the substrate may accumulate inside the cell in concentrations much higher than the external concentrations (Figure 11). However, if energy coupling is blocked or energy generation stops, the concentrating uptake no longer occurs because entry and exit of the substrate occur through the membrane at the same rate by facilitated diffusion only. Therefore, active transport mechanisms permit intracellular accumulation of substrates which exist in low external concentrations by the expenditure of energy to overcome the concentration gradient. The internal concentration becomes maximal and further transport may cease, thus showing a saturation effect on this active transport system. With the facilitated diffusion system, on the other hand, the uptake rate and the intracellular level are proportional to the external concentration just as in a simple diffusion system (Figure 12).

By another mechanism, the substrate is converted into a derivative as part of the process of translocation across the membrane. This process is called group translocation, whereby the substrate passes across the membrane as a chemical group. As the product by this process is chemically different from the external substrate, no concentration gradient is produced across the membrane. For instance, the sugars transported through group translocation are phosphorylated by the phosphotransferase system;

i.e., a certain sugar being transported is phosphorylated by a small heat-stable protein designated HPr, which acts as a carrier of high-energy phosphate. Then the phosphorylated sugar moves into the cell.

These processes in the active transport system, therefore, require the performance of work. Both the facilitated diffusion and group translocation process apparently are linked to the accumulation against a concentration gradient, showing dependence on Michaelis-Menten kinetics at low substrate concentrations because of carrier involvement, just like enzyme involvement in biochemical reactions. Coupling of energy-generating systems to the transport system permits the concentrative uptake of substrates, whereby the concentration inside the cell may reach as much as 100 to 10,000 times the external concentration. These transport systems are found extensively in procaryotes, and similar mechanisms also are found in eucaryotes. Transport mechanisms performed by mitochondria in eucaryotes are similar to those of cell membranes in procaryotes. Even the transport mechanisms of the cell membrane in certain eucaryotes have been shown to be similar to those in procaryotes. For instance, Hellebust[158] has shown that one species of marine diatom, *Melosira nummuloides*, has the ability to take up amino acids through an active transport system.

Many microorganisms are able to use a very wide range of natural material. However, only low molecular weight chemical compounds can penetrate the microbial cell. High molecular weight chemicals, such as nucleic acids and polysaccharides, are hydrolyzed by exoenzymes into soluble sizes that can pass through the cell membrane.

Once passing through the cell membrane the chemical compounds are directly available for oxidation by the electron transport system. Electron transport systems are attached to or are an integral part of some membranes, because all cytochromes are found in membranes. In procaryotes, the electron transport system is associated with the plasma membrane system where catabolic reactions take place. In eucaryotic organisms, on the other hand, the cytochromes and electron system are associated with the mitochondria, and thus many catabolic reactions are located in the mitochondria and microbodies. Since most aquatic microorganisms are aerobic, they can oxidize organic compounds by respiration. Respirable compounds usually can be oxidized through the TCA cycle after conversion of each compound into pyruvate or acetyl-CoA. Hydrocarbons, fatty acids, and many alcohols are oxidized only by respiration and not by fermentation.

Many ATP molecules are produced by substrate-level phosphorylation during the oxidation-reduction processes. It also is possible to obtain some of the energy released during an oxidation-reduction process called electron-transport phosphorylation; the energy released by the transfer of electron coenzymes involved in oxidation-reduction reactions to O_2 is conserved through synthesis of ATP. These oxidation-reduction reactions involve coenzymes such as nicotinamide adenosine dinucleotide (NAD), nicotinamide adenosine dinucleotide phosphate (NADP), and flavin adenine dinucleotide (FAD). The $NADH_2$ or $NADPH_2$ can be oxidized to NAD or NADP through the electron-transport system, producing three ATP molecules per molecule of $NADH_2$ or $NADPH_2$ oxidized. Two molecules of ATP are produced per molecule of FAD oxidized.

When heterotrophic uptake of organic substrate is performed by the carrier-mediated transport processes in aquatic environments with low substrate concentrations, the substrate undergoing the transport (S) combines with a carrier protein (C), as

$$S + C \rightleftharpoons CS$$
(substrate) (carrier) (carrier-substrate complex)

This process is similar to the biochemical process of

$$
\underset{\text{(enzyme)}}{\text{E}} \ + \ \underset{\text{(substrate)}}{\text{S}} \ \rightleftharpoons \ \underset{\text{(enzyme-substrate complex)}}{\text{ES}} \ \rightarrow \ \underset{\text{(enzyme)}}{\text{E}} \ + \ \underset{\text{(product)}}{\text{P}}
$$

where E represents the enzyme combining with the substrate S to give an intermediate complex ES that breaks down into the products P and the enzyme. Therefore, the kinetics of nutrient uptake can be described using Michaelis-Menten enzyme kinetics as:

$$
v = \frac{V_m S}{K_m + S}
$$

where v is the rate of substrate uptake, V_m is the maximum rate of substrate uptake, K_m is the substrate concentration at which $v = V_m/2$ and S is the substrate concentration. The constant K_m is an important property of a microbial cell because it reflects the ability of microorganisms to utilize a substrate down to a certain threshold concentration. The cell of a species such as a marine bacterioplankton has a very low K_m value, showing saturation effect at low external concentration of a substrate. Thus the heterotrophic uptake of soluble organic compounds in natural water can be explained in terms of Michaelis-Menten enzyme kinetics.[25] In most natural waters the active transport of substrate is effective at concentrations below 100 $\mu g/\ell$. On the other hand, the simple or facilitated diffusion of substrate through the cell membrane occurs at concentrations above 500 $\mu g/\ell$ and has been attributed to eucaryotic phytoplankters.[159] In many aquatic ecosystems the major fraction of organic materials is soluble. The living fraction of such ecosystems, which has a relatively small biomass, contributes to or influences the total fraction only slightly.[32] The total amount of organic solutes in the hydrosphere may be very large, but the concentration is generally too low to serve as a major source of nutrient for most aquatic organisms. This in turn leaves organic solutes relatively free from direct assimilation by aquatic communities, and thereby the ecosystem can be stable. Thus, in natural waters, the active uptake of substrates by bacteria generally keeps the concentration of every substrate dissolved in water below 20 μg C/ℓ. Based on the theory that the active transport system for heterotrophic uptake follows the original expression given by Michaelis and Menten, the dynamics of dissolved organic materials in natural waters can be measured by employing the radioactive tracer technique.

The uptake kinetics of organic substances by microorganisms in the surface waters of oligotrophic marine regions may be characterized by those in the southern (Station 9) and northern regions (Station 11) of the subtropical convergence within the western North Pacific central water, in the Kuroshio Current (Station 12) and in the subarctic Pacific water (Station 19) of the Pacific Ocean (Figure 13A). The lower boundaries of the surface water masses of the southern and northern regions of the western North Pacific central water, the Kuroshio Current, and the subarctic Pacific water were found during the investigation at depths of approximately 100, 100, 100, and 50 m, respectively.

A substrate specificity of substrates examined was apparent in the uptake kinetics of microorganisms in each water mass (Table 3). The turnover time for the dissolved organic substances by the microorganisms was several months, both in the southern and northern regions within the western North Pacific central water. It was also shown that the subtropical convergence in the western North Pacific central water has little effect on the regional difference between the uptake kinetics of microorganisms in seawater of the southern and northern regions. On the other hand, the turnover time was a few thousand hours in the Kuroshio Current and the subarctic Pacific water, whereas several to ten thousand hours was required in the western North Pacific central water. This observation supports a general assumption that the turnover rate of biological elements in both temperate and subarctic regions may be greater than that in the subtropical region of the ocean. This is based on the fact that the rate in the subtropical region was several times slower than in temperate and subarctic regions. No significant difference in the uptake kinetics of each substrate was apparent within most of each surface water mass. In oligotrophic waters the patchiness of microbial biomass and activities may cause serious problems in biological investigations. However, when each sample size was a few liters in volume no differences were observed in both the sample from a euphotic (center of the water mass: 50 m) and an aphotic (bottom of the water mass and the nitrite maximum layer: 100 m) location in the surface layer of the western North Pacific central water (Tables 3 and 4). A relatively homogeneous distribution of uptake kinetics as well as of the population density of microorganisms and the *in situ* substrate concentrations were observed. Therefore, when the average value of each parameter is taken on the scale of a few liters then vertical mixing of seawater obviously is sufficient to overcome large-scale vertical zonation.

Generally, eutrophication has progressed in coastal regions of the marine environment. Many marine water masses in these regions show characteristics similar to those of mesotrophic or sometimes eutrophic lakes. Mesotrophic characteristics usually are observed in the coastal regions of open ocean, whereas eutrophic features can be observed in semienclosed bays, such as Tokyo Bay, or in inland seas, such as the Adriatic Sea and the Seto Inland Sea in Japan, where large amounts of municipal and industrial wastes have been discharged.

During the summer stagnation period, the water masses at Station T-1 in Tokyo Bay (Figure 13B) are vertically stratified into three layers (also see Chapter 2, Figure 14): surface, intermediate, and bottom.[161] The uptake kinetics of the dissolved organic substances in these water masses are shown in Table 5. This table also shows the kinetics in the surface water mass of the plume off the Edo-gawa River (Station T-2) where both municipal and industrial wastes have been discharged. The turnover times of the dissolved organic substrates were within a few days and usually between 10 and 24 hr. The turnover rate was high in the surface and bottom layers, but low in the intermediate layer. *In situ* uptake rate of each substrate was more than half of the maximum attainable rate of uptake for almost all the samples in such a hypereutrophic environment, thereby the *in situ* concentrations of these substrates in Tokyo Bay during the summer stagnation period were high enough to support almost optimal activities of microorganisms in the seawater. In every sample *in situ* substrate concentration of protein hydrolysate, an amino acid mixture, was shown by bioassay to be almost the same amount as the sum of the individual amino acid substrates examined, i.e., aspartic acid, glutamic acid, glycine, alanine, and lysine. The fraction of mineralization in the gross assimilation of each substrate by microorganisms in almost every sample at *in situ* substrate concentration was from 20 to 30% for most of the amino acids and usually more than 50% for carbohydrates.

As an example of eutrophication in the coastal waters of open ocean, a large amount of municipal wastes have been discharged into Shimoda Bay without the influence of industrial wastes. During the summer stagnation period the surface water mass of Shimoda Bay is influenced greatly by the freshwater inflow of the Inohzawa River, and it runs off into the Kuroshio Counter Current which also is influenced by a branch of the Kuroshio Current (Figures 13A and B). The water mass in Shimoda Bay (Station S-2) is identical with an eutrophic one, and both those of the Kuroshio Counter Current (Station S-3) and a branch of the Kuroshio Current (Station S-4) are identical with mesotrophic ones. The concentration of each substrate (Sn) was shown by the bioassay derived from the determination of uptake kinetics (Tables 6 to 9) to be kept usually below 20 $\mu g/\ell$ in seawater, even in such an eutrophic water mass as that in Shimoda Bay. This concentration is the threshold kept by the active transport of most substrates by microorganisms in oligotrophic waters.[32,163] Thus microorganisms, even in eutrophic water masses, must react very quickly to an increase of a substrate concentration by keeping the amount of dissolved organic carbon at approximately 1000 μg C/ℓ near the surface of any water mass. The turnover times of most dissolved organic substances were within a few days in both the Inohzawa River and Shimoda Bay where freshwater microorganisms as well as marine microorganisms contribute to the uptake kinetics. The turnover time was a few hundred hours in the Kuroshio Counter Current or in the branch of the Kuroshio Current (Tables 10 and 11); *in situ* uptake rate of each substrate was less than one tenth of the maximum attainable rate of uptake in the same water masses or of the *in situ* uptake rate of Shimoda Bay. The fraction of mineralization in gross assimilation of each substrate by microorganisms at *in situ* substrate concentration, both in the Kuroshio Counter Current and the branch of the Kuroshio Current, was generally between 20 and 30% for most of the amino acids and 50% for the carbohydrates.

In the marine environment, therefore, the critical turnover time of easily metaboliz-able substances dissolved in seawater seems to lie between a few months and a few weeks for oligotrophic and mesotrophic levels, and between a few weeks and a few days for mesotrophic and eutrophic levels.

In contrast with most of the rivers in Japan, many rivers in Canada discharge into the Pacific Ocean without appreciable influences of municipal and industrial wastes. For instance the surface water of the Stamp River (Figure 14) is characterized by being low in both standing stock of microorganisms and in primary productivity. The turn-over time of dissolved organic substances in the Stamp River ranged from a few days to a few weeks with a relatively high turnover rate of 62 hr for glycolic acid (Table 12). Accordingly, the water of the Stamp River may be characterized as meso-oligo-trophic from the uptake kinetics of the organic substances by microorganisms. This characterization is identical with that determined by physiochemical or biological pa-rameters.[165] The Stamp River is connected at the upper reaches with Great Central Lake (Figure 14), where commercial fertilizer (nitrogen and phosphorus) has been added to increase primary and secondary production and thus to boost the survival and abundance of under-yearling sockeye salmon (*Oncorhynchus nerka*).[166-173] The biological and chemical parameters at both unfertilized and fertilized regions in Great Central Lake are still within the steady-state oscillations of a temperate oligotrophic lake in spite of weekly surface additions of nitrogen and phosphorus.[165,169] Despite these weekly additions of chemical fertilizers during the summer stagnation period, inorganic nutrient concentrations have been low in the euphotic zone indicating rapid utilization by phytoplankton. The higher primary productivity of the fertilized region compared to the unfertilized region has not been reflected in a higher phytoplankton standing stock. Although phytoplankton growth is highest in the region of greater nutrient availability, advection of the biomass in the surface layer of the lake must be occurring. The turnover rates of the dissolved organic substances were mostly within a few weeks (Tables 13 and 14). During an initial period (Table 14), significant activity occurs in the surface layer of the lake with a relatively rapid assimilation of carbohy-drates and organic acids, substrates mainly used by microorganisms for their energy generation systems. The assimilation of amino acids, mainly used for their structural systems, is slow. This is expected in a stable oligotrophic aquatic ecosystem since the active catabolism of amino acids by heterotrophic microorganisms does not efficiently link the food chain together and sometimes produces toxic organic nitrogen excretions. The responses of heterotrophic microorganisms buffer ecosystem impacts, and it is apparent that fertilization has not upset the heterotrophic feedback mechanisms expected in an oligotrophic lake. By the controlled addition of inorganic nutrients there has been a direct increase in phytoplankton production and an indirect increase in the microbial biomass serving as particulate food for filter-feeding zooplankton and for the mineralization of organic substances and recycling of biological elements.

One example of ultraoligotrophic waters in the freshwater environment may be glacially oligotrophic lakes. In Meziadin Lake and Kitlope Lake (Figure 14), both typical examples of glacial lakes, nitrogen is present in excess and phosphorus is present in very low concentrations. The common features of rapid flushing, low base nutrient load, and weak stratification with deep mixing throughout the year have contributed to the ultraoligotrophic status. *In situ* concentrations of amino acids and carbohydrates (Table 15) in these lakes are at an order of magnitude lower than those in eutrophic water masses,[162] but organic acids are at relatively high concentrations. In the surface water of the slightly glacial Meziadin Lake, the turnover time of the dissolved organic substances was similar to rates in ultraoligotrophic marine waters, but turnover time in the moderately glacial Kitlope Lake was equivalent to those of meso-oligotrophic marine waters (Table 15). Turnover rates of organic acids were much higher than those of amino acids and carbohydrates in the glacial lakes.

Another unique freshwater environment is the dystrophic type of lake. There, as for instance in dystrophic Lowe Lake and Bonilla Lake (Figure 14), concentrations of both nitrogen and phosphorus are low and distinct summer depletion of epilimnetic nitrogen occurs, reflecting low total chlorophyll concentrations, low bacterial density, and low primary production. The turnover time of dissolved organic substances in the moderately humic Bonilla Lake was shown to be similar to those of mesotrophic marine waters, and that in the slightly humic Lowe Lake it was equivalent to that in meso-oligotrophic marine waters (Table 16). The turnover rates of organic acids were almost the same or slower than those of amino acids and carbohydrates in the dystrophic lakes. The mineralization fraction of gross assimilation for organic acids by microorganisms is almost the same as that for amino acids in the dystrophic lakes where high concentrations of humic organic materials, largely of terrestrial plant origin, are present.

Water of the Stamp River flows finally into Alberni Inlet (Figure 14). A large amount of organic materials is discharged by the pulp mill at the head of Alberni Inlet, and therefore heterotrophic microorganisms predominate over phytoplankton even in the euphotic zone at the head and central region of the inlet. In the surface water at the entrance of Alberni Inlet the turnover times of dissolved organic substances range from several days to a few weeks (Table 17). These slow turnover rates are due to the predominance of autotrophic organisms over heterotrophic organisms in the water. The turnover times inside Alberni Inlet, on the other hand, were considerably less and ranged from several hours to several days; the bacterial concentration was only slightly higher. The *in situ* concentrations of each dissolved substrate in the surface waters at the entrance and the central region of the inlet are usually within the range of those measured in the surface water masses of the Canadian coast.

As all of these uptake kinetics have been examined during the optimal season for microbial activities, the turnover times of easily metabolizable organic solutes have been shown to be characterized by different thresholds in oligotrophic, mesotrophic, and eutrophic water masses as shown clearly in Figure 15 for glutamic acid, as an example. The turnover time in each type of trophic water mass seems to have a steady-state oscillation within a certain range that is bounded by the thresholds. This phenomenon may be true not only during the optimal season but may hold without regard to seasonal changes, at least for glutamic acid in the eutrophic water mass of Shimoda Bay and its neighboring mesotrophic water mass of the Kuroshio Counter Current

(Figure 16). Each threshold, however, is not exactly the same at every boundary between the same trophic water masses. This is especially evident in the differences of the freshwater and marine environments. Accordingly, this apparent paradox possibly may be resolved using Ohle's theory,[176] which shows that input of a nonlimiting element profoundly affects uptake of trace elements by modification of the chemical dynamics. Amplitude of the oscillation increases within the range according to enrichment of a water mass, and finally the system in any water mass may reach a higher trophic state by irreversibly crossing over a threshold.

Temperature was shown not to be the primary factor in regulating the steady-state oscillation, at least in the water mass of Shimoda Bay, as temperature dependence of the turnover rate for every substrate examined by the Arrhenius plot was not obvious (Figure 17). Further, the oscillation could not be regulated primarily by salinity, pH, carbon dioxide, or phosphate. However, it can be regulated greatly by inorganic nitrogen (Figure 18). The oscillation, as indicated by the relation of turnover time of amino acids and the concentration of inorganic nitrogen, shows a definite U-shape for every substrate of amino acid; i.e., the turnover rate is lower at both borders of the range of inorganic nitrogen concentrations. As inorganic nitrogen is the product of a catabolic pathway involving amino acids, the mechanism of the steady-state oscillation must be controlled chiefly by a negative feedback system; i.e., when inorganic nitrogen in the system is low in concentration, induction of active uptake of substrates (amino acids in this case) occurs and their mineralization is accelerated. When excess amounts of the products (inorganic nitrogen in this case) are built up in the environment, the product in the catabolic pathway inhibits the active uptake of substrates and their mineralization. This control of the steady-state oscillation of the kinetics in a certain water mass could be comparable to that of feedback inhibition in an in vivo biochemical pathway, where the activity of the first enzyme of the pathway is inhibited by the end product and thus controls the production of the end product. In this way, inorganic nitrogen (the end product in the mineralization pathway of amino acids) inhibits the activity of a membrane-bound carrier. However, when the inorganic nitrogen in the environment is depleted by phytoplankton below a certain level, active transport can resume, just as a biochemical reaction can resume when the end product is depleted in vivo. Hence the steady-state equilibrium of concentration of each organic substance can be maintained in an aquatic ecosystem on the level of molecular biology. This cybernetic system may be revealed as the mechanism controlling oscillation of turnover time of dissolved organic substances in the euphotic zone by acting as a negative feedback system at the community level, primarily through the biological activities of phytoplankton and bacteria. A product in a certain catabolic pathway inhibits active uptake systems of microorganisms when excess amounts of the products are built up in the environment. Such a product is generally an inorganic compound and is used up by phytoplankton. The phytoplankton synthesize organic compounds and supply them, partly in dissolved forms, to the heterotrophic microorganisms. Then the active transport systems for uptake of these substrates by the microorganisms can resume their functions.

By such a cybernetic system as described above, each turnover time of an organic substance in a water mass of any trophic state seems to maintain a steady-state oscillation within a certain range that may be bounded by thresholds. According to the degree of eutrophication of a water mass, amplitude of the oscillation increases within the range, and the system in the water mass finally reaches the next trophic state by irreversibly crossing over the threshold. The turnover time of easily metabolizable organic substances thus has been estimated by Wright-Hobbie uptake kinetics to be less than a few days in hyperentrophic waters, a few days in eutrophic waters, and between a few days and several tens of days in mesotrophic waters, and between a few months and a few hundred years in oligotrophic waters.[160-162,165,175,178-182]

In the dysphotic or aphotic zones where phytoplankton cannot carry on photosynthesis, the detritus food chain takes precedence over the grazing food chain. The stability of the steady-state equilibrium of organic substances can be maintained in these zones only by the interaction among the communities comprising the detritus food chain, otherwise the balance of nature in an ecosystem may be destroyed, followed by a serial destabilization of the steady-state equilibrium of some important environmental factors.

Creating experimental disturbance by changing the influx of energy or materials *in situ* can induce microbial reactions for readjustment to maintain a steady-state oscillation of uptake kinetics of dissolved organic substances. Two different levels in the flow rates of energy and material were performed by setting up a large subsurface enclosure (controlled enclosure bag: diameter 5 m; height from the sea floor 17.5 m) *in situ* in Patricia Bay, Saanich Inlet, B.C., Canada (Figure 14). This device (Figure 19) eliminates active photosynthesis of phytoplankton inside the bag in the topmost layer of 2 to 5 m depth depending on the tide, and thereby the influxes of energy and organic material from the euphotic zone to the sea floor are different inside and outside the bag.

The ecosystem in the dysphotic zone of Patricia Bay can generally be categorized as a moderately mesotrophic system as derived from the euphotic zone. When precisely analyzed, however, the dynamics of dissolved organic substances in the dysphotic zone seem to be a little different from those in the euphotic zone; i.e., in the dysphotic zone turnover rates of amino acids were higher while those of carbohydrates and organic acids were lower (Tables 18 and 19). This could be due to the higher turnover rates of organic compounds which are connected more directly to the TCA cycle in the euphotic zone. Hence the rate of influx into the dysphotic zone of chemical compounds more closely connected with the TCA cycle should be lower. The difference in dynamics also can be attributed to differences in the species composition of microbial communities in euphotic and dysphotic zones. In the dysphotic zone the majority of organic materials may be supplied from the euphotic zone and consumed by heterotrophic communities comprising the detritus food chains.

Immediately after the influence of primary production on the dysphotic communities was experimentally reduced by the enclosure of the water mass within the enclosure bag, heterotrophic activities became very active (i.e., within the range of a mesotrophic system for a few days). Then these activities decreased but still remained within the range of a mesotrophic system. Thereafter the system tended to recover to the original steady-state equilibrium of a moderately mesotrophic system (Figure 20). In this case, the organic substances dissolved in the seawater oscillated essentially in a sine curve showing an equilibrium between input (chiefly supplied from phytoplankton in the euphotic zone) and output (chiefly assimilated by heterotrophic microorganisms). The oscillation may be modified by a change of organic supply from the euphotic zone. Therefore, as an example of the dynamics of easily metabolizable organic substances, the best regression for the turnover time of glutamic acid at 1 m above the sea floor inside the bag (Figure 20) may be simulated as:

$$\text{Turnover time of glutamic acid (hrs)} = \left[300 \sin\left\{\frac{\pi}{27}(t+10)\right\}\right] \exp\left\{\frac{-(t-50)^2}{3000}\right\} + $$

$$\left[180 \sin\left\{\frac{\pi}{105}(t+5)\right\}\right] \exp\left\{\frac{-(t-350)^2}{20000}\right\} + 280$$

where t is time (hr) elapsed since the initiation of the experiment at 11 a.m. on July 17, 1980. The wavelength for the oscillation of the turnover time of glutamic acid was 27 hr, and the turnover time converges theoretically at 280 hr. During the experiment, glutamic acid and organic carbon dissolved in the seawater inside the bag were maintained at average concentrations of 6.8 ± 4.6 µg/ℓ and 1.59 ± 0.28 mg C/ℓ, respectively.

On the other hand, the best regression for the turnover time of glutamic acid in the natural environment at 1 m above the sea floor outside the bag (Figure 21) may be simulated as:

$$\text{Turnover time of glutamic acid (hrs)} = \left[100 \sin\left\{ \frac{\pi}{25}(t+5) \right\} \right] \exp\left\{ \frac{-(t-80)^2}{3000} \right\} +$$

$$50 \exp\left\{ \frac{-(t-230)^2}{8000} \right\} + 100$$

The wavelength for the oscillation of the turnover time of glutamic acid was 25 hr, and the turnover time converges theoretically at 100 hr. The exponential stress in this simulated model was due primarily to a large supply of organic substances from a phytoplankton bloom in the euphotic zone that occurred just after the experiment was initiated. However the fluctuation of the turnover time of glutamic acid does not seem to be affected greatly by the exponential element and to be almost exclusively in a steady-state oscillation. Incidentally, the average concentrations of dissolved organic carbon and of glutamic acid were 1.95 ± 0.20 mg C/ℓ and 5.4 ± 4.4 µg/ℓ, respectively. This concentration of glutamic acid outside the bag was not significantly different from that inside the bag. The concentration of dissolved organic carbon, on the other hand, was significantly higher outside than inside the bag. The turnover time inside the bag was almost three times greater than that under natural conditions, but the steady-state oscillation was still moderately mesotrophic although toward the oligotrophic side. As these two equations based on observations inside and outside the bag could be expressed in the same fundamental wave, the time lag of these two phenomenona can be calculated to be 5 hr. Therefore, the indirect influence of the enclosure (as indocated primarily by the decrease in the indicated influx of organic substance) is to retard the response of the uptake kinetics of organic solute by the microbial communities in the marine environment, as revealed by the turnover retardation of organic substances. Moreover, the microbial readjustment to a new balance after the impact was done by a negative feedback system reflected by a fluctuation in the sine curve that still kept the environment within the mesotrophic level. This steady-state oscillation maintained by a negative feedback system, therefore, is similar to a classical prey-predator relationship. A phenomena which enhances stability of the biosphere may be present in any ecological system.

The population densities of bacterioplankters, which are primarily responsible for the utilization of organic solutes, also have been shown to be in steady-state oscillation within each water mass, both in freshwater and marine environments (Figure 22). The range of this oscillation seems to be very approximately divided by the thresholds among oligotrophic, mesotrophic, and eutrophic water masses. The roughness in this division may be reasonable since the population densities of bacterioplankton are on a community level, whereas the uptake kinetics of organic substances are on the more precise level of molecular biology.

Aritifical enrichment contributes greatly to the eutrophication of aquatic ecosystems, and that may be one of the greatest impacts of human activities on the delicate equilibrium of the biosphere. The appearance of man as a member of the community in the biosphere has not only resulted in local or temporary changes in the delicate equilibrium of the biosphere, but has also threatened to alter the total balance of the cycle of matter on Earth. Human activities, however, have not yet caused such perturbation on the environment in terms of scale and speed as those brought about by major upheavals in the history of the Earth. Before the mid-1800s when the populations of towns increased as a result of the industrial revolution between 1760 and 1830, human population in the world was small and sewage disposal presented few problems. Although the Greeks and Romans had hygienic systems, the discharge of sewage seems to have had little effect on water pollution in natural environments. In many oriental countries, lavatory wastes were removed by farmers to nightsoil reservoirs in paddy fields for later fertilization of the fields (Figure 23). In other cases, lavatories were constructed over fish cultivation ponds, whereby the human refuse was used as food for omnivorous fish cultivated there. In these instances, the discharge of human activities was used to enhance food production. During the Middle Ages and Renaissance in Europe, human habitations resembled pig sties, with steps and odd corners being used as lavatories and chamber pots being emptied into the streets (Figure 24). As unsanitary as it was from the standpoint of public health, however, the biological elements involved in human society were cycled in semienclosed, centralized towns and villages. In an extreme case resulting from urbanization, the River Thames by 1860 had become a vast open sewer, with the rain-washed refuse of London flowing into the river. Thus domestic wastes can cause serious eutrophication, not only in the freshwater environment but also in coastal regions of the marine environment. Sewage disposal began to be carried out partly to maintain sanitary standards and partly to prevent the contamination of rivers and quiet waters with organic materials. The disposal of organic wastes has become a very important problem in the course of industrial development and consequent concentration of human populations in big cities.

Agriculture also causes serious eutrophication in certain locations of the aquatic environment. Lake Kasumigaura in Japan (Figure 25) is one such environment which is artificially enriched chiefly by agricultural drainage from paddy fields and pig farming. The particles in the water at such an extreme stage of eutrophication are primarily comprised of biogenic organic matter. Phytoplankters are major components of the biogenic materials throughout the year, especially during the blue-green algal bloom (Figure 26).[187] During the exponential growth phase of blue-green algae, the ATP percentage in the biomass may be almost one order of magnitude greater than that during the death phase (Figure 27). Heterotrophic microorganisms may comprise almost the same biomass, and their concentration fluctuates by one or two orders of magnitude compared to that of phytoplankters throughout the year. The abundance of microorganisms is related to the growth phase of phytoplankters, particularly that of the blue-green algae, as can be seen in the case of bacterioplankton (Figure 28). Thus bacterial multiplication seems to be greatly depressed in the exponential growth phase of blue-green algae and to be stimulated by their death phase. Due to such enhancement of heterotrophic processes over autotrophic processes according to the degree of eutrophication, the fraction of particulate organic matter increases firstly in heterotrophic microorganisms, secondly in phytoplankton, and finally in detritus (Figure 29). Therefore, total organic matter in natural waters can be shown to approximate the following distributions in relative units:

	Oligotrophic waters[38,39,70]	Mesotrophic waters[162]	Eutrophic waters[37,186-188]	Hypereutrophic waters[37]
Organic solutes	100	100	100	100
Detritus particles	1	3	3	200
Phytoplankton	2	20	20	2000
Bacteria and allied microorganisms	0.2	4	2	60

During initial enrichment of a water mass in the oligotrophic state the concentration of dissolved organic matter increases but there is no appreciable increase in particulate organic matter. When the system in the water is enhanced up to the mesotrophic state both dissolved and particulate organic materials increase according to the enrichment. Finally the concentration of dissolved organic matter attains a plateau and does not increase appreciably, even with additional enrichment, but this does increase the concentration of particulate organic matter within the system of a eutrophic stage. When a eutrophic system is enriched even more, the dissolved fraction of organic matter seems to be transferred into the particulate fraction by heterotrophic bacteria exclusively. The bacteria involved in this transformation are not oligotrophic species, such as *Caulobacter* and *Spirillum lunatum*, but others that can be cultured by a 0.1% peptone medium. Every chemical compound in particulate form has a much slower biodegradation rate than in the dissolved form. Moreover, maintaining the concentration of dissolved organic matter in any natural water below the threshold of facilitated diffusion in the nutrient uptake systems of living organisms restricts availability of these compounds to only those organisms which have active transport systems. Accordingly, the precedence of heterotrophic processes over autotrophic processes must favor not only detrital food chains but also the stability of an ecosystem in maintaining a steady-state type, such as oligotrophic or eutrophic; otherwise the excess production of organic matter leads the system into disequilibrium and results in a change in water type.

Easily metabolizable organic materials are not the sole organic materials that flow into aquatic environments from terrestrial environments. Among organic compounds that cannot be readily decomposed are some chemicals present in wood, petroleum, asphalt, polythene, and possibly all kinds of poisonous substances such as insecticides. Many aquatic microorganisms, especially marine bacteria, have been shown to be able not only to tolerate low concentrations of these toxic substances but also to slowly decompose them and grow at their expense. This biochemical versatility of microorganisms may be greater than that of higher organisms.[189] However, all creatures in the biosphere share basic biochemical properties although a few properties may differ from species to species; i.e., each organism shares the same cellular building blocks of organic substances, the same major biochemical processes, the same control systems in catabolic and anabolic processes, and the same energy generation systems. All biological elements, therefore, undergo cyclical changes based on the same kinetic principles, even in various ecosystems of the biosphere.

III. CYCLES OF BIOLOGICAL ELEMENTS

Elements such as carbon, nitrogen, oxygen, phosphorus, sulfur, hydrogen, iron, magnesium, and silicon comprise the structure of biological molecules and undergo comparable cyclical changes. At least six major biological elements are contained in biological substances in the biosphere: hydrogen (H), carbon (C), oxygen (O), nitrogen (N), phosphorus (P), and sulfur (S). For example, $(CH_2O)_{106}(NH_3)_{16}H_3PO_4$ has been used as the average chemical model of plankters in lower trophic levels of the marine ecosystem.[190] Another well-known example in a higher marine trophic level is meal protein of cod that has the formula $C_{265}H_{555}O_{174}N_{83}S$. These elements undergo cyclical transformation from inorganic combinations to organic combinations and back again by biological activities. The maintenance and creation of new life on the Earth is absolutely dependent on the cyclic turnover of these elements. When solar energy is trapped and converted to chemical bond energy by photosynthetic organisms, high energy phosphate bonds are synthesized. Hence phosphorus is of primary importance since it is coupled with the first stage of energy transmission. Since ATP is short-lived

and the energy in ATP is quickly used in the conversion into the chemical bond energy
of carbon, phosphorus undergoes rapid cycling in vivo and thus is not required in large
quantities. However, carbon is concerned in the large quantities of the chemical bond
energy accumulated in the various organic constituents. This progression always is
involved with those changes involving hydrogen and oxygen (Figure 30).

A. Carbon Cycle

Carbon makes up 0.02% of the total elementary composition of the crust of the
Earth, including the hydrosphere (0.0028%) and atmosphere. The majority of carbon
is present as carbon dioxide in the form of bicarbonate and carbonate ions, undissociated molecules of carbon dioxide and carbonic acid. The total amount of carbon
dioxide in the marine environment has been estimated to be 1.3×10^{20} g. Nearly 70%
occurs in carbonate, a little in carbon dioxide, approximately 25% in organic compounds, 0.03% in fossil fuels and 0.015% in biomass and undecayed remains of organisms. Most living creatures contain approximately 40% carbon on a dry weight basis.

The oxidized forms of carbon (carbon dioxide, bicarbonate, carbonate) are converted to reduced forms of carbon (organic substances) through autotrophic processes.
Photosynthesis may be the major pathway of the autotrophic processes on the scale
of the whole biosphere, but the contribution of chemosynthesis and heterotrophic
processes may be appreciable in many locations of the aquatic environment. The dark
fixation of carbon dioxide through chemosynthetic and heterotrophic processes has
been commonly detected, not only at certain specific sites such as the boundary at
approximately 150 m between the aerobic surface layer and anaerobic layer in the
Black Sea where chemosynthesis can be prominent over other autotrophic pathways
(Figure 31), but also at the interface between the bottom sediment and its overlying
water in any natural water in shallow areas (e.g., in the marine environment,[42] in the
freshwater environment,[ ] and in the deep ocean.[54,129,193,194] Just as in the Black Sea,[192]
environmental conditions in these latter locations are changing from aerobic to anaerobic and chemosynthetic processes can be predominant. In a typical coastal region of
Japan, for instance, seasonal changes can lead to the dark fixation of carbon dioxide
at the sediment surface of Aburatsubo Inlet (35° 09′ N, 139° 37′ E) being as high as
20% of the photosynthesis throughout the year.[42] The dark uptake in the water column
after the spring bloom of phytoplankton can amount to the same rate as that of photosynthesis (Figure 32). In aerobic environments, the subsequent oxidation of these
organic products through heterotrophic processes is coupled with the reduction of molecular oxygen back to water. Through both autotrophic and heterotrophic processes,
therefore, the cyclic transformations of carbon and oxygen are dependently linked to
each other, whereby carbon dioxide has been believed to be in balance on a global
scale. Both low concentrations of carbon dioxide and high concentrations of oxygen
gas can be limiting to photosynthesis, and this favors stability of the steady-state equilibrium of carbon dioxide concentration in the atmosphere. This balance, however, is
shifted slightly in the direction of an increase in the atmosphere (Figure 33) through
human activities which release a tremendous amount of carbon dioxide from the burn-
ing of the fossil fuel pool. As the photosynthetic zone in the terrestrial environment is
now decreasing, also by human activities, readjustment of the increase of carbon dioxide may depend exclusively on the capacity of the carbonate system in the marine
environment, which is very efficient in removing carbon dioxide from the atmosphere.
This process, however, may have a certain impact on the chemical balance in the marine ecosystem.

B. Nitrogen Cycle

Nitrogen makes up 0.002% of the total elementary composition of the crust of the
Earth, including the hydrosphere (0.00005%) and atmosphere. More than 99.99% of

the nitrogen in the atmosphere is N_2 which is the principal reservoir of nitrogen in the biosphere. Seawater of the open ocean has a nitrogen content of about 0.5 ppm. Approximately 62% of the nitrogen in the marine environment occurs as nitrate, 0.003% as nitrite, 0.007% as ammonium, and approximately 37% as organic nitrogen. Most living organisms contain 9 to 15% nitrogen on a dry weight basis. All organisms depend on chemical transformations for their energy, and these involve the oxidation and reduction of chemical compounds. In those redox transformations relevant to energy generation systems, nitrogen plays a complicated role because it has a large number of oxidation levels. In highly reduced states, nitrogen exists either in the form of the ammonium ion or of amino compounds. In highly oxidized states, nitrogen is in the form of nitrate ions; i.e., to convert nitrogen in ammonium ions or amino acids to nitrogen in nitrates involves a total valence change of eight, and vice versa. The nitrogen atoms combine with other elements to form protein molecules. These protein molecules constitute the major part of most living organisms, amounting to tens of thousands of atoms. These proteins can be involved in diverse biochemical functions, and thereby the nitrogen cycle in the biosphere is regulated by biological activity. Inorganic nitrogen, in the form of ammonia, nitrites or nitrates, is relatively scarce in aquatic environments. Nitrogen concentration becomes the limiting factor in some water masses in the aquatic environment, as can be explained by Liebig's law of the minimum, although phosphorus may be the most common limiting nutrient in natural waters.

While nitrate or nitrite is the form in which nitrogen is assimilated by aquatic organisms, it must be reduced in their cells to ammonium and then incorporated into the amino group ($R-NH_2$) through the activity of glutamic dehydrogenase. These organic nitrogenous compounds are synthesized by plants and microorganisms; thereafter they are utilized as nitrogen sources for animals. The nitrogenous compounds assimilated by animals remain at almost the same reduced levels as those in plants and microorganisms.

The form of nitrogen excreted as waste products varies from one group of animals to another. Invertebrates excrete predominantly ammonia. Reptiles and birds excrete predominantly uric acid, and mammals excrete mostly urea. In nature, the excreted urea and uric acid are utilized rapidly by microorganisms with the formation of ammonia. This process of ammonification also takes place through the hydrolysis of plant or animal remains. Dead organic constituents immediately are attacked by microorganisms, and the nitrogenous compounds are decomposed with liberation of ammonia. Part of the nitrogen is converted into microbial cell constituents. The conversion of ammonia to nitrate is called nitrification, and this conversion is brought about in water by two groups of obligately aerobic, chemolithotrophic bacteria in two steps: ammonia is oxidized to nitrite by *Nitrosomonas*; then nitrite is oxidized to nitrate by *Nitrobacter*. On the other hand, the conversion of nitrate into nitrogen gases under anaerobic conditions is called denitrification. Denitrifying bacteria, such as *Pseudomonas denitrificans*, can release the nitrogen of nitrates as free nitrogen molecules, resulting in a loss of nitrogen available for most members of the community in the biosphere. This loss of biological nitrogen is compensated for by the activity of nitrogen-fixing microorganisms. The principal groups of free-living, nitrogen-fixing microorganisms in aquatic environments are certain blue-green algae such as *Anabaena spiroides*, bacteria belonging to the genus *Azotobacter*, and certain anaerobic bacteria of the genus *Clostridium*. Nitrogen gas is reduced first to ammonium by this nitrogen fixation process. Then ammonium is converted into organic form. Nitrogen molecules (N_2) are extremely inert and their reduction requires high energy. The reduction process is catalyzed by the enzyme nitrogenase. These nitrogen-fixing microorganisms have been shown to be of primary importance in the nitrogen fixation process of the nitrogen cycle, although artificial fertilizers, thunderstorms, and ultraviolet light from the sun may fix nitrogen to a limited extent without the intervention of biochemical reactions. In the biosphere, 1000 million tons of nitrogen are estimated to pass into the nitrogen cycle every year (Figure 34).

On the basis of the features of inorganic nitrogen metabolism, Koike and Hattori[196,197] have classified coastal sediments into three types:

1. Muddy sediments rich in organic materials with restricted oxygen supply, where anaerobic metabolism and denitrification predominate
2. Sandy sediments rich in oxygen supply which allows nitrifying bacteria to flourish and thus the nitrification process to predominate
3. Sediments having intermediate characteristics of the above two types, where denitrification and nitrification occur simultaneously

In this third type of sediment, denitrification accounts for a certain fraction of the nitrate reduction. With respect to the first type, in any muddy sediment, irrespective of the location in the coastal region of Japan, the rate of nitrogen gas production in the sediment surface has been shown to be in the order of 10^{-2} μg atom N per g sediment per hr.[196,197]

C. Phosphorus Cycle

Phosphorus makes up 0.105% of the total elementary composition of the crust of the Earth, including the hydrosphere. It comprises 0.000007% of seawater. The vast majority of phosphorus in aquatic environments is present as inorganic phosphates.

The primary energy carrier in biological systems is ATP. The high energy bonds of ATP comprise two of the phosphate bonds. ATP is used as the primary energy carrier in a wide variety of biochemical reactions because the synthesis and hydrolysis of ATP is independent of oxidation-reduction potentials in biological systems. Hence, no life is possible without phosphorus, and all organisms contain phosphorus in amounts ranging from 0.05% in certain plants to as much as 6% in certain vertebrates, on a dry weight basis. Most phosphorus occurs in cells as organophosphates (nucleotides, nucleic acids, phospholipids, phosphoproteins, etc.) with ionic phosphate. Microorganisms and many animals decompose organophosphorus compounds in dead material with the liberation of phosphate. Such phosphate is assimilated by plants and heterotrophic microorganisms for their phosphorus requirements.

In aquatic environments, microbial activities play an important part in the solubilization and precipitation of phosphates. Phosphates precipitate in high pH waters and are soluble in low pH waters. Most phosphate precipitating in the marine environment is in the form of $Ca_3(PO_4)_2$. In certain euphotic zones primary production by phytoplankton is restricted by a shortage of available phosphates. In other euphotic zones of coastal regions or inland waters, extensive algal blooms develop and form red tides because of an excess supply of phosphates. Generally in most oligotrophic waters phosphorus is recognized as the limiting nutrient. In eutrophication of a water mass (Figure 35), there is a definite correlation between the degree of enrichment and two important factors, phosphate nutrient loading and mean depth of the water.[198]

D. Sulfur Cycle

Sulfur makes up 0.025% of the total elementary composition of the crust of the Earth. It makes up 0.0885% in seawater. Many microorganisms can utilize inorganic sulfate (SO_4^{2-}) as the sole source of sulfur to synthesize amino acids and other sulfur-containing compounds. The assimilated sulfate is reduced to sulfite (SO_3^{2-}), then to sulfide (HS^-) and finally to organic sulfur by reaction with serine to synthesize cysteine.

Other organic sulfur compounds can be further synthesized from cysteine. Sulfate-reducing bacteria belonging to the genus *Desulfovibrio* reduce sulfate as a terminal electron acceptor by reducing hydrogen sulfide (H_2S) as the end product. Because seawater contains a relatively high concentration of sulfate, sulfate reduction is a common process in the mineralization of organic materials on the sea floor. Sulfur also is liberated from the sea floor into overlying water in the reduced inorganic form of hydrogen sulfide; this occurs when organic sulfur compounds in the dead bodies of plants and animals are mineralized in the sediment. The hydrogen sulfide, however, does not accumulate in seawater, except under anaerobic conditions, because it is rapidly and spontaneously oxidized in the presence of dissolved oxygen gas. The biological oxidation of hydrogen sulfide and of elemental sulfur is brought about by autotrophic bacteria which require oxygen. Reduced sulfur can be oxidized either aerobically by the colorless sulfur bacteria (*Thiobacillus, Thioploca, Thiothrix, Thiospirillum*) or anaerobically by the photosynthetic purple and green sulfur bacteria (*Chromatium, Chlorobium, Rhodothiospirillum, Thiopedia*).

The cycle of each biological element in the biosphere involves the constant turnover of all elements being geared together in a combined and integrated operation. The integration of these various cycles results in a steady-state equilibrium balanced by production and consumption of biologically important materials in the biosphere as it exists at present. Accordingly, if only one cycle breaks down, all of these cycles are destroyed.

D. Turnover Rates of Organic Materials

Heterotrophic bacteria can utilize a wide variety of organic compounds in natural waters. The constituents of these compounds may be divided into three broad categories with special reference to biodegradation: (1), constituents easily metabolizable by most microorganisms such as amino acids, monosaccharides and organic acids; (2), constituents moderately resistant to biochemical breakdown such as cellulose and chitin; and (3), refractory constituents highly resistant to biochemical breakdown such as aquatic humus.

Firstly, the turnover times of easily metabolizable constituents have already been shown to be less than a few days in hypereutrophic waters, a few days in eutrophic waters, between a few days and several tens of days in mesotrophic waters, and several tens of days in the surface layer of oligotrophic waters. Secondly, the turnover times of moderately resistant constituents have been shown to be between several days and several tens of days in hypereutrophic waters, between several tens of days and several months in eutrophic waters, several months in mesotrophic waters, and a few years in oligotrophic surface waters. Finally, the decomposition rates of most refractory constituents of organic materials may be regulating the outside cycling rim of turnover of biological elements as a whole.

Actually, in the shallow photosynthetic layer of most natural waters, total production or organic matter in the water column exceeds mineralization for most of the year. This may be compensated for by either the laying down of rich organic sediments or by the transportation of organic materials out of the immediate environment by advective effects.[42] When this budget is analyzed in each enclosed system, however, more than 99% of these organic materials produced in aquatic environments are decomposed at an appreciable rate. The remaining fraction, known as humus, is highly resistant to biochemical decomposition. Although humus was defined originally as organic matter which is resistant to bacterial decomposition, in fact humus is not absolutely resistant. It partly consists of residues of plants and animals autochthonous in the aquatic environment, although some aquatic humus is of terrestrial origin. Therefore, the slowest cycling of organic substances may take place with humus and other refractory organic materials.

The dynamics of biochemically refractory substances are different in various water masses of different degrees of eutrophication. In Lake Kasumigaura, as an extreme example of eutrophication, the standing stocks of dissolved organic matter and of the bacterioplankton which depend solely on dissolved matter for nutrition oscillated in a complicated manner (Figure 36), but the fundamental wave in the oscillation must be driven through the formation of dissolved organic matter by phytoplankton.[187] The oscillation in concentration of dissolved organic matter and abundance of bacterio-plankton may be approximated by a sine curve, as there should be an equilibrium between input, chiefly supplied from phytoplankton, and output, chiefly assimilated by bacterioplankton. Even though the organic supply from a phytoplankton bloom might accumulate during the early stages of the bloom, the bacterioplankton then increase in abundance by utilizing the nutrients. Thereafter, an increase in bacterial density leads to an active consumption of the organic matter and the release of inorganic nutrients which become available for the formation of the next phytoplankton bloom. A series of these processes forms one cycle of the sine curve, and such cycles may be repeated many times in a year with amplitudes that essentially depend on the size of the phytoplankton bloom. Thus bacterial utilization of dissolved organic matter has been examined with respect to four seasonal periods of phytoplankton succession (Figures 37 to 40).[37]

The standing stock of dissolved organic matter increased through the production of phytoplankton in summer (Figure 38) and spring (Figure 40), and it decreased in early summer through the decomposition processes of bacterioplankton after the spring bloom of phytoplankton (Figure 37) and also in autumn and winter after the summer bloom of phytoplankton (Figure 39).

The decomposition rate of dissolved organic matter (dDOC/dt) and the multiplication rate of bacterioplankton (dTB/dt) can be defined by a differential equation modelled on the standing stock. Thus, the decomposition rate of dissolved organic carbon (DOC) during the blue-green algal bloom in summer was

$$\frac{dDOC}{dt} = 0.0251 \cos\left[\frac{\pi}{60}(t-78.5)\right] \times 10^{\left[0.577 + 0.479 \sin\frac{\pi}{60}(t-78.5)\right]}$$

where t is days since May 1, 1978. The multiplication rate of bacterioplankton was

$$\frac{dTB}{dt} = 0.0230 \cos\left[\frac{\pi}{60}(t-87.2)\right] \times 10^{\left[13.339 + 0.444 \sin\frac{\pi}{60}(t-87.2)\right]}$$

The fluctuation of dissolved organic carbon in the summer bloom, therefore, preceded that of the bacterioplankton by 8.7 days (Figure 38).

During the eucaryotic phytoplankton bloom in spring the decomposition rate of dissolved organic carbon was

$$\frac{dDOC}{dt} = -0.0101 \cos\left[\frac{\pi}{48}(t-280.6)\right] \times 10^{\left[0.277 - 0.155 \sin\frac{\pi}{48}(t-280.6)\right]}$$

The multiplication rate of bacterioplankton was

$$\frac{dTB}{dt} = -0.0312 \cos\left[\frac{\pi}{48}(t-263.9)\right] \times 10^{\left[12.853 - 0.476 \sin\frac{\pi}{48}(t-263.9)\right]}$$

There was a lag time of 14 days between dissolved organic carbon in the spring bloom and the change in population level of bacterioplankton (Figure 40).

During the period between these blooms, rates of the dissolved organic carbon decomposition and the bacterial multiplication were not highly dynamic. Rates after the spring bloom (Figure 37) were

$$\frac{dDOC}{dt} = -0.00538 \cos\left[\frac{\pi}{36}(t-133)\right] \times 10^{\left[0.443 - 0.0617 \sin \frac{\pi}{36}(t-13.3)\right]}$$

and

$$\frac{dTB}{dt} = 0.0179 \cos\left[\frac{\pi}{36}(t-0.725)\right] \times 10^{\left[13.11 + 0.205 \sin \frac{\pi}{36}(t-0.725)\right]}$$

and, after the summer bloom (Figure 39), were

$$\frac{dDOC}{dt} = 0.0188 \cos\left[\frac{\pi}{45}(t-201.40)\right] \times 10^{\left[-0.00983 + 0.270 \sin \frac{\pi}{45}(t-201.4)\right]}$$

and

$$\frac{dTB}{dt} = 0.0150 \cos\left[\frac{\pi}{45}(t-187.2)\right] \times 10^{\left[12.40 + 0.215 \sin \frac{\pi}{45}(t-187.2)\right]}$$

The turnover rate of dissolved organic matter represents the time required for the organic matter to be entirely removed by the heterotrophs; it can be determined as the standing stock of dissolved organic carbon divided by its decomposition rate. Accordingly, the turnover time of dissolved organic matter can be calculated to be 52 days during the blue-green algal bloom in summer, 61 days after the summer bloom in autumn, 101 days during the eucaryotic phytoplankton bloom in spring, and 188 days after the spring bloom in early summer. These estimates of the turnover times of organic solutes in the lake include decomposition of almost all chemical compounds, even such biochemically stable substances as aquatic humus, because the dissolved organic carbon concentration was as low as 100 $\mu g/\ell$ at the end of autumn when heterotrophic activity was the greatest for the year. Such low concentrations already have been measured, even in oligotrophic waters of the deep ocean.[199] The many particles associated with numerous microorganisms in the water might have the same activity as activated sludge produced in the aerobic treatment systems of sewage and waste waters. Incidentally, the turnover time of aquatic humus in this hypereutrophic water is a few hundred times slower than that of easily metabolizable substances, such as amino acids in waters of the same trophic stage.[112,164,177]

In an ultraoligotrophic oceanic water mass, taken as the opposite extreme example of eutrophication, the dynamics of biochemically stable organic substances are inactive. The kinetics of the steady-state equilibrium in such cases have been well described by Olson[200] for terrestrial environments. In the simplest case, when there is a continual input (L) to a standing stock of organic carbon (X), the change in X with time can be described by the equation:

$$X = \frac{L}{k}(1 - e^{-kt})$$

where k represents the rate (time^{-1}) at which the standing stock is being decomposed. In a situation in which sufficient time elapses (i.e., t is large enough), the system will reach a steady-state value, X_{ss}, which is equal to L/k.

Skopintsev[201] has estimated the steady-state value of dissolved organic carbon in deep oceanic waters. The total amount of dissolved organic carbon (X) in the oceans may be 1.3×10^{18} g, assuming an average concentration of dissolved organic carbon in oceanic waters of 1.0 g C/m^3 [202] and a total volume of oceanic waters of 1.3×10^{18} m^3. The annual input of autochthonous dissolved organic carbon (L) will be 11.5×10^{14} g C/year for the whole volume of the ocean, since the annual input of aquatic humus of planktonic origin is 3% of the annual primary production. The annual input of allochthonous humus of terrestrial origin is 1.8×10^{14} g C/year. The input of organic materials from the atmosphere into the ocean is small. The annual input of aquatic humus into the ocean, therefore, totals 13.3×10^{14} g C/year. Accordingly, the residence time may be calculated as:

$$\frac{1.3 \times 10^{18} \text{ g}}{1.33 \times 10^{15} \text{ g/year}} = 977 \text{ years}$$

This residence time can be expressed as the time required for the aquatic humus to reach 95% of the steady-state value of Olson.[200] This value, which is given by 3/k, corresponds to 2900 years. This estimated time requirement to reach 95% of the steady-state value of aquatic dissolved humus in the ocean is similar to that of Williams et al.,[203] who used radioactive carbon dating and estimated the apparent age of dissolved organic materials from deep waters (1800 and 1920 m depth) of the northwest Pacific Ocean to be 3740 ± 300 years.

On the other hand, a major fraction of organic materials is transported from the surface layer into the deep ocean in the form of rapidly sinking large particles such as fecal pellets of zooplankton.[136] As an example of the turnover rate of particulate organic materials in shallow coastal regions, the turnover time of such particles has been determined in Departure Bay (49° 12.63′ N, 123° 57.28′ W) in British Columbia, Canada[31] by examining the balance between the amount of organic material entering the dysphotic system and the amount retained there, as estimated by the theory of Olson.[200] Using a sediment trap, the amount of particulate organic carbon sedimented during a 1-year period (L) was measured to be 200 g/m^2. The quantity of organic carbon (X_{ss}) beneath 1 m^2 was 2010 ± 330 g in the immediate area of the sediment. This value is assumed to approximate the steady-state carbon content of the sea floor. Then the decomposition rate factor (k) can be calculated to be 0.10. Thus the regeneration rate of particulate organic carbon can be estimated at 30 years, which is the time required for the organic material to reach 95% of the steady-state value in the shallow sea floor.

The same approach can be used to estimate the turnover rate of particulate organic carbon in the deep ocean. As an example of turnover rate of particulate organic material in deep oceanic regions, the turnover rate has been determined on the ocean floor in the northern North Pacific Ocean (47° 51.1′ N, 176° 20.6′ E), using data of Tanoue and Handa.[114] Using a sediment trap, the amount of particulate organic carbon sedimented during 1 year (L) was measured as 0.896 g/m^2. The quantity of organic carbon (X_{ss}) beneath 1 m^2 was 893 g in the immediate area of the sediment. Then the decomposition rate factor (k) was calculated to be 0.000973. Thus the decomposition rate of particulate organic carbon can be expressed as 3083 years, or the time required for the organic material to reach 95% of the steady-state value in the deep ocean floor. This turnover rate of particulate organic material in the deep ocean is almost the same value as those calculated by Skopintsev[201] and Williams et al.[203] for dissolved organic materials.

When the two values in Departure Bay and in the deep ocean of the northern North Pacific Ocean are compared, the rate of bacterial activities may be 10 to 100 times slower in the deep ocean than in the shallow sea. Therefore, the turnover rates of all categories of organic compounds may be one or two orders of magnitude higher in the shallow layer than the deeper layer, and thereby the turnover times of the easily metabolizable constituents and of those constituents which are moderately resistant to biochemical breakdown may be between a few months and a few years and a few tens of years, respectively, in the deep ocean.

Therefore the turnover times of refractory constituents, both in dissolved and particulate form, can be estimated to be between half a year and one year in hypereutrophic waters, several years in eutrophic waters, several tens of years in mesotrophic waters, between several tens of years and hundreds of years in the surface layer of oligotrophic waters, and several thousands of years in the deep layer of oligotrophic waters (Table 20).

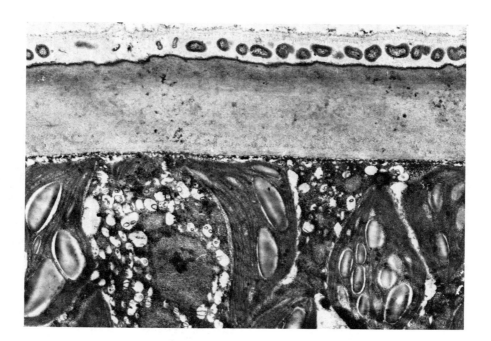

FIGURE 1.   Bacteria living on the cell surface of a green alga, *Bryopsis maxima*. (Courtesy of Dr. T. Hori.)

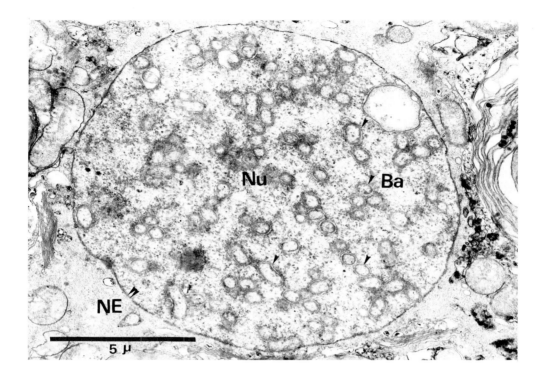

FIGURE 2. Bacteria living inside a living cell of *Chattonella* species. Ba, bacteria; Nu, nucleus; NE, nuclear membrane. (Courtesy of Dr. Y. Hara.)

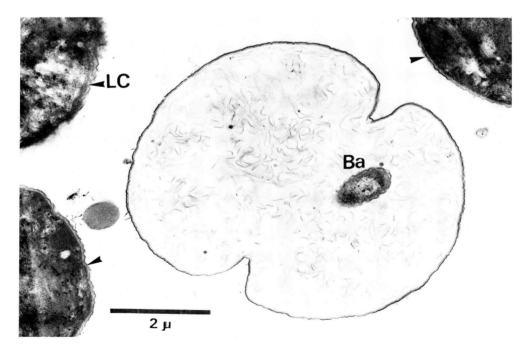

FIGURE 3. Bacterial participation in the degradation of a dead cell of *Microcystis aeruginosa* in its early stage of autolysis. Ba, bacteria; LC, living cell of *M. aeruginosa.* (Courtesy of Dr. Y. Hara.)

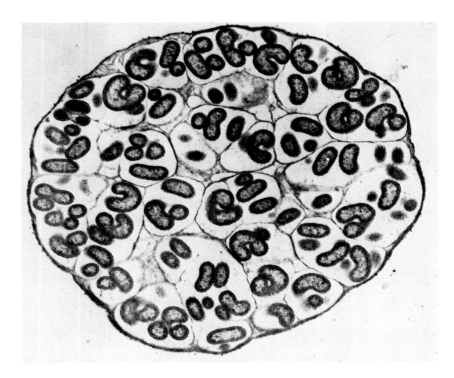

FIGURE 4.   Bacteria multiplying in the matrix of dead protoplasm pushed out from a dead cell of a prasinophyte species, *Bipedinomonas rotunda*. (Courtesy of Dr. T. Hori.)

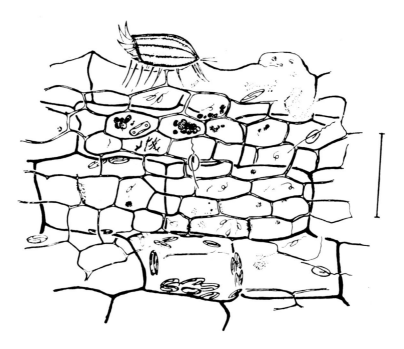

FIGURE 5.   The microbial community of a detrital particle derived from a *Thalassia* leaf. Scale: 0.1 mm. (From Fenchel, T., *Limnol. Oceanogr.*, 15, 14, 1970. With permission.)

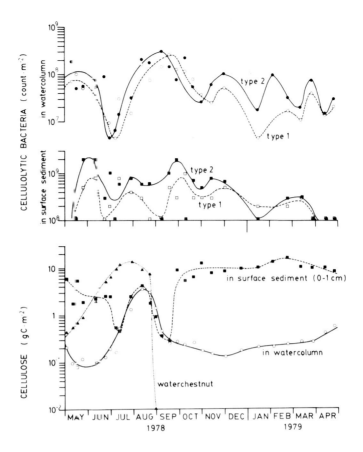

FIGURE 6   Cellulose dynamics in the water chestnut ecosystem. Cellulolytic bacteria Type 1 can utilize cellulose as its sole energy source. Cellulolytic bacteria Type 2 cannot utilize cellulose as its sole energy source but excretes cellulase. (From Matsuo, S., Yamamoto, H., Nakano H., and Seki, H., *Water Air Soil Pollut.*, 12, 511, 1979. With permission.)

FIGURE 7.   A heavy blue-green algal bloom completely covers the leaves of water chestnut.

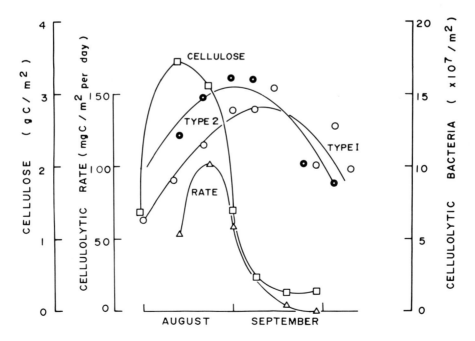

FIGURE 8.   Simulated model of cellulose dynamics in the water chestnut ecosystem of Lake Kasumigaura, Japan. □ Standing stock of cellulose; △ cellulolytic rate; ○ cellulolytic bacteria Type 1; ● cellulolytic bacteria Type 2.

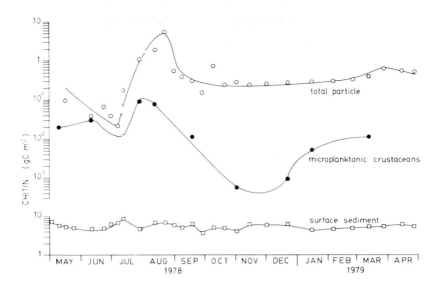

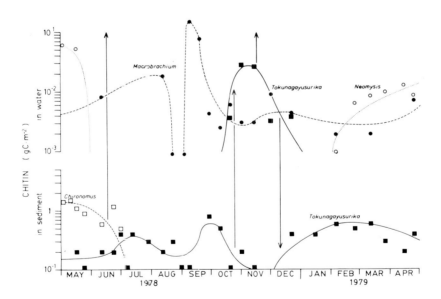

FIGURE 9. Seasonal fluctuations of chitin in biomass of predominant arthropods and detritus particles (From Yamamoto, H. and Seki, H., *Water Air Soil Pollut.*, 12, 519, 1979. With permission.)

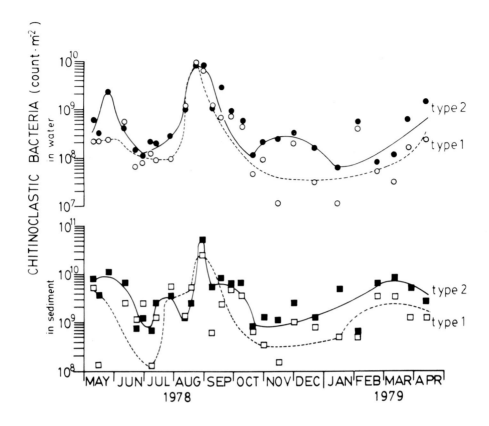

FIGURE 10.   Seasonal fluctuation of the population density of chitinoclastic bacteria. (From Yamamoto, H. and Seki, H., *Water Air Soil Pollut.*, 12, 519, 1979. With permission.)

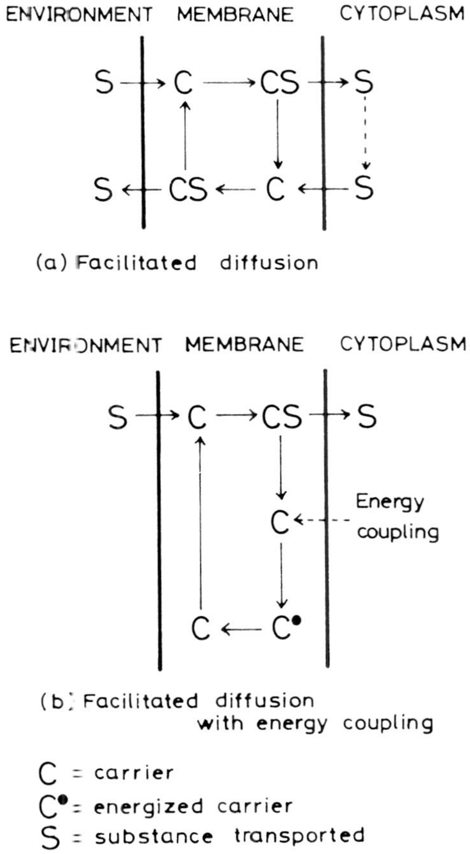

ENVIRONMENT MEMBRANE CYTOPLASM

(a) Facilitated diffusion

ENVIRONMENT MEMBRANE CYTOPLASM

Energy
coupling

(b) Facilitated diffusion
with energy coupling

C = carrier
C• = energized carrier
S = substance transported

FIGURE 11.   Model for functions of carrier in facilitated diffusion. (a) At equilibrium, exit balances entrance. Transport occurs but accumulation against a concentration gradient does not. (b) Energized carrier
has greatly reduced affinity for carrier on the cytoplasmic side of the membrane. Exit is much reduced over
entrance. Accumulation against a concentration gradient occurs. In neither case does the carrier move lat-
erally in the membrane, but remains in one location and undergoes changes in conformation. (From Brock,
T. D., *Biology of Microorganisms*, Prentice-Hall, Englewood Cliffs, N. J., 1979, 802. With permission.)

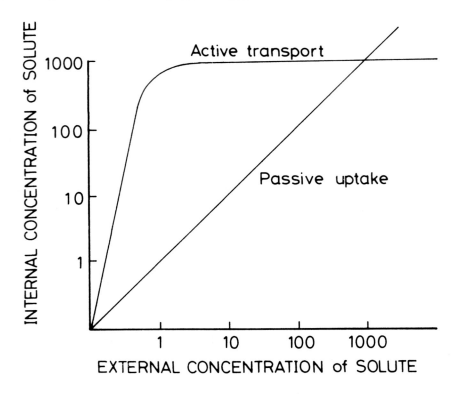

FIGURE 12.   Relationship between external and internal solute concentrations in passive uptake and active transport. Note that in passive uptake the external and internal concentrations are identical. In active transport the internal concentration is higher than the external concentration but shows saturation at high external concentrations (From Brock, T. D., *Biology of Microorganisms*, Prentice-Hall, Englewood Cliffs, N.J., 1979, 802. With permission.)

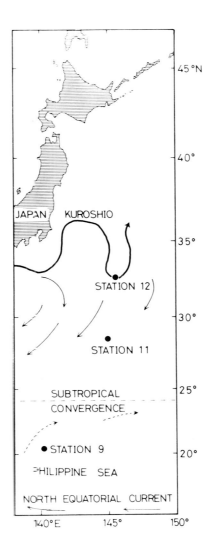

FIGURE 13A.   Hydrography and station locations in the western North
Pacific central water, the Kuroshio Current, and the subarctic Pacific
water of the Pacific Ocean.

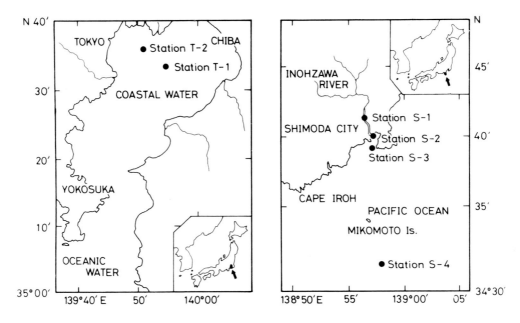

FIGURE 13B.   Station locations in Tokyo Bay and Shimoda Bay of the coastal regions of Japan.

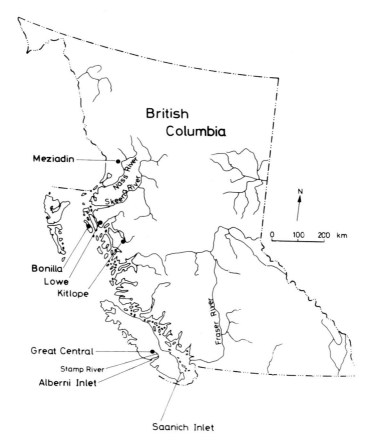

FIGURE 14.   Location of the Stamp River, Great Central Lake, Meziadin Lake, Kitlope Lake, Lowe Lake, Bonilla Lake, Alberni Inlet, and Saanich Inlet in British Columbia, Canada.

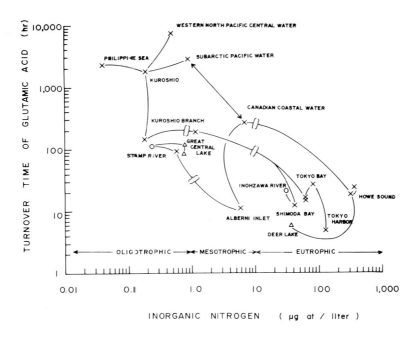

FIGURE 15. Turnover time of glutamic acid in different trophic water masses of the Pacific region during the optimal season for microbial activities. (From Seki, H., *Proc. 15th Pac. Sci. Congr.*, in press. With permission.)

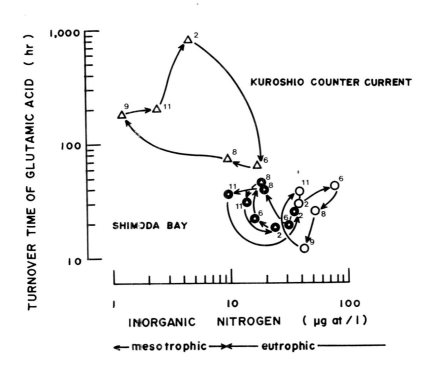

FIGURE 16. The mode of steady-state oscillation as indicated by the relation of the turnover time of glutamic acid and concentration of inorganic nitrogen in water of Shimoda Bay and the Kuroshio Counter Current. (From Seki, H., Terada, T., and Ichimura, S., *Arch. Hydrobiol.*, 88, 219, 1980. With permission.)

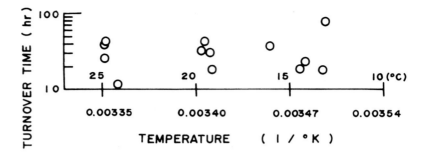

FIGURE 17.   The Arrhenius plot for the turnover time of glutamic acid in water of Shimoda Bay. (From Seki, H., Terada, T., and Ichimura, S., *Arch. Hydrobiol.*, 88, 219. 1980. With permission.)

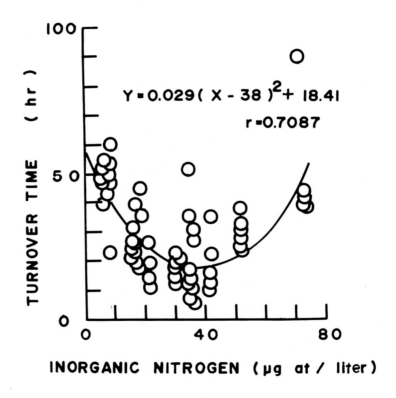

FIGURE 18.   Steady-state oscillation in the relation of turnover rate of amino acids and the concentration of inorganic nitrogen in the water mass of Shimoda Bay. On the turnover rate of amino acids (glutamic acid, glycine, alanine, aspartic acid, and lysine), the effect of substrate specificity (amino acids) was not significant ($F = 0.0505$, $F_{0.05} = 2.63$) but the effect of mineralized product, inorganic nitrogen, was highly significant ($F = 7.22$**, $F_{0.01} = 2.94$).[177] (From Seki, H., *Proc. 15th Pac. Sci. Cong.*, in press. With permission.)

135

BOTTOM MOUNTED COMPONENT

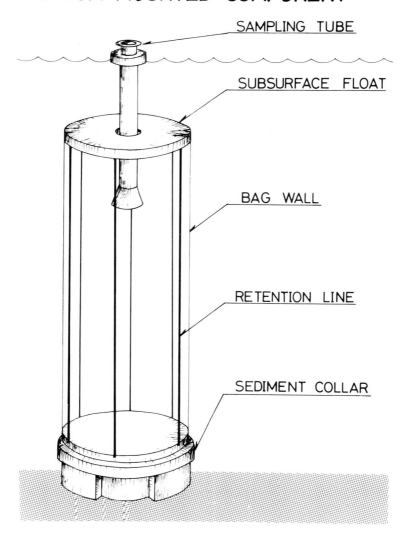

FIGURE 19.   Enclosure bag launched *in situ* in Patricia Bay at a depth of approximately 20 m.

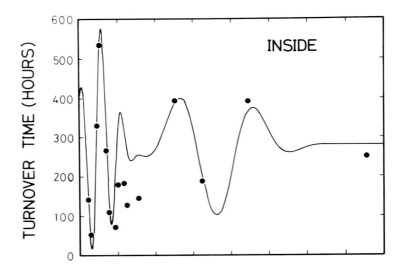

FIGURE 20.   The turnover time of glutamic acid at 1 m above the sea floor in the dysphotic waters inside enclosure bag launched in Patricia Bay. (From Seki, H., Aoshima, N., Whitney, F., and Wong, C. S., *Water Air Soil Pollut.*, in press. With permission.)

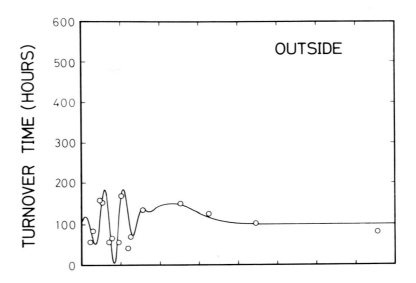

FIGURE 21.   The turnover time of glutamic acid at 1 m above the sea floor in the dysphotic waters outside enclosure bag launched in Patricia Bay. (From Seki, H., Aoshima, N., Whitney, F., and Wong, C. S., *Water Air Soil Pollut.*, in press. With permission.)

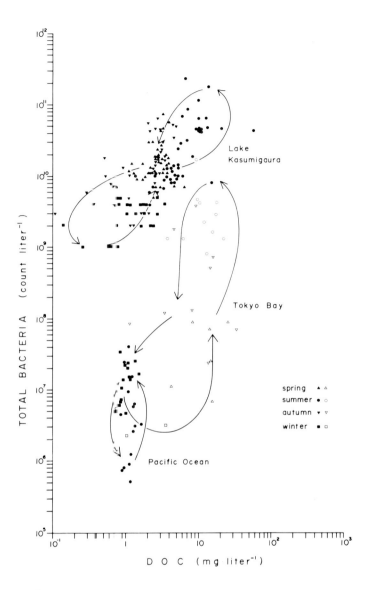

FIGURE 22. Relationship between standing stock of bacterioplankton and dissolved organic carbon in various natural waters. Bacterioplankton, enumerated in a bacterial counting chamber; DOC, determined by TOC Analyzer Model 915 B (Beckman, Fullerton), followed by the filtration of water sample using Gelman glass filter type A (pore size, 0.3 μm). (From Seki, H. and Nakano, H., *Kieler Meeresforsch.*, in press. With permission.)

FIGURE 23.   A night soil reservoir (↓) in a paddy field. The organic components of feces are mineralized by microbial activities in the reservoir into inorganic nutrients which then are used for enrichment of surrounding paddy fields. From a part of the woodcut print of Hamamatsu scenery in ''Old-Time Fifty Three Stages on the Tokaido Highway'' by Ando Hiroshige (1797—1858).

FIGURE 24.   A German print showing the low public health standard of making babies relieve nature over the street.

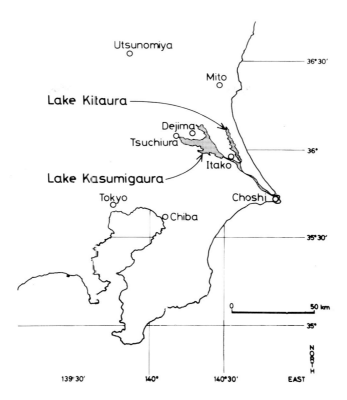

FIGURE 25.   Location of Lake Kasumigaura, a hypereutrophic lake
in Japan. Takahama-iri Bay exists at the head of the northern part of
the lake where water chestnut (*Trapa bispinosa*) has expanded its distribution onto the littoral shelves.

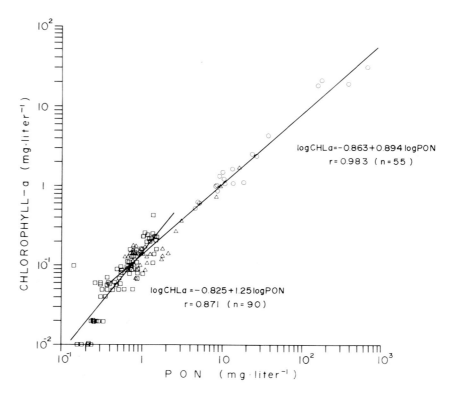

FIGURE 26.   Relationship between particulate organic nitrogen and chlorophyll *a* in the lake water of Lake Kasumigaura, Japan. o, During the period of heavy bloom of blue-green algae; Δ, in the early stage of the blue-green algal bloom; □, during the period without the formation of blue-green algae. (From Nakano, H. and Seki, H., *Water Air Soil Pollut.*, 15, 215, 1981. With permission.)

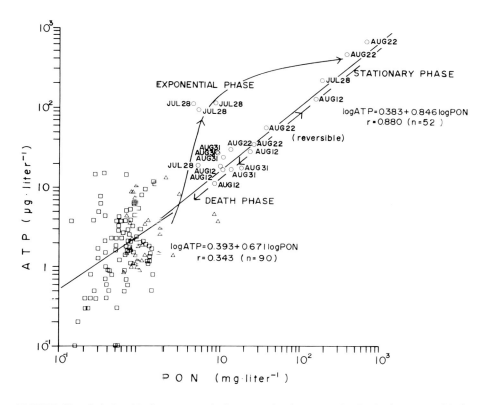

FIGURE 27. Relationship between particulate organic nitrogen and ATP in the water of Lake Kasumigaura, Japan. Symbols are the same as in Figure 26. (From Nakano, H., and Seki, H., *Water Air Soil Pollut.*, 15, 215, 1981. With permission.)

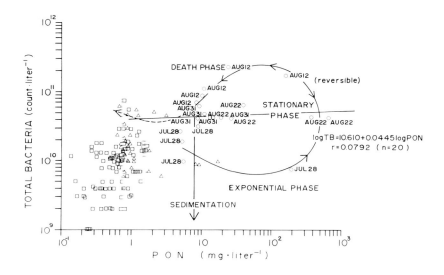

FIGURE 28. Relationship between particulate organic nitrogen and total number of bacterioplankton. Symbols are the same as in Figure 26. (From Nakano, H. and Seki, H., *Water Air Soil Pollut.*, 15, 215, 1981. With permission.)

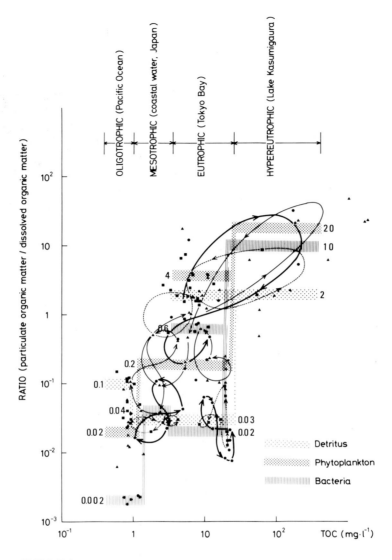

FIGURE 29.   Constituent distribution of organic matter in various aquatic environments with special reference to eutrophication. (From Seki, H. and Nakano, H., *Kieler Meeresforsch.*, in press. With permission.)

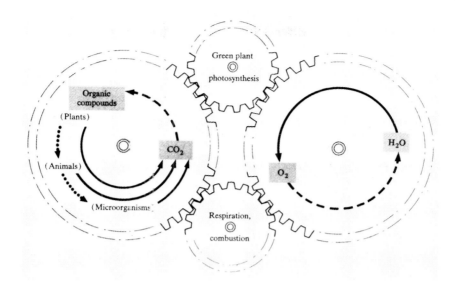

FIGURE 30. The carbon and oxygen cycles. The oxidations of the carbon and oxygen atoms are shown as solid arrows. The reductions are shown as broken arrows. Reactions involving no valence change are shown as dotted arrows. (From Stanier, R. Y., Doudoroff, M., and Adelberg, E. A., *The Microbial World*, Prentice-Hall, Englewood Cliffs, N.J., 1963, 753. With permission.)

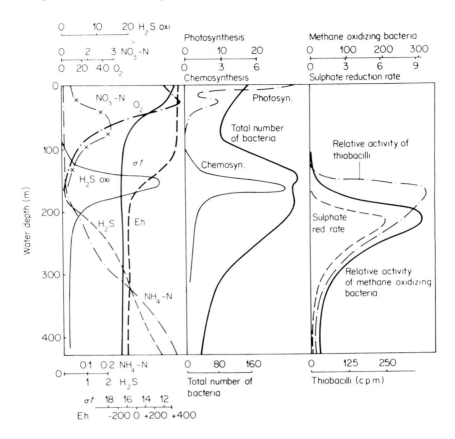

FIGURE 31.   Chemistry and microbial activities on a vertical profile in the Black Sea. O_2 (% Saturation, mg/ℓ); σ_t; Eh (mV); NO_3-N (mg/ℓ); NH_4-N (mg/ℓ); H_2S oxidation rate (μg/ℓ/day); photosynthesis (mg C/m^3/day); chemosynthesis (mg C/m^3/day); total number of bacteria ($\times$ 10^3/mℓ); relative activity of methane oxidizing bacteria (cpm/mℓ); relative activity of thiobacilli (cpm/mℓ); rate of sulfate reduction (μg S^{--}/ℓ/day). (From Sorokin, Y. I., *Marine Ecology*, Kinne, O., Ed., John Wiley & Sons, New York, 1978, 501. With permission.)

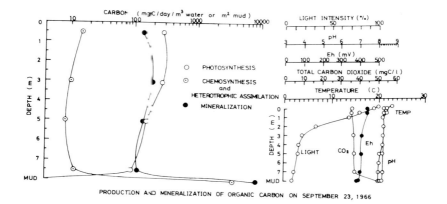

PRODUCTION AND MINERALIZATION OF ORGANIC CARBON ON SEPTEMBER 23, 1966

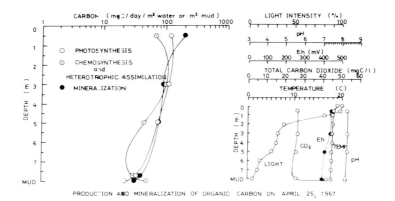

PRODUCTION AND MINERALIZATION OF ORGANIC CARBON ON APRIL 25, 1967

FIGURE 32. Carbon budget in Aburatsubo Inlet at two different times of the year. (From Seki, H., *J. Fish. Res. Bd. Can.*, 25, 625, 1968. With permission.)

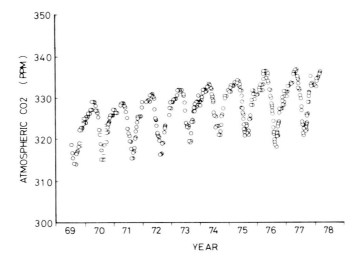

FIGURE 33. The increase and oscillation of carbon dioxide in the atmosphere at Station P (50° 00′ N, 145° 00′ W) in the northeastern subarctic Pacific Ocean. (From Wong, C. S. *Marine Pollut. Bull.*, 9, 264, 1978. With permission.)

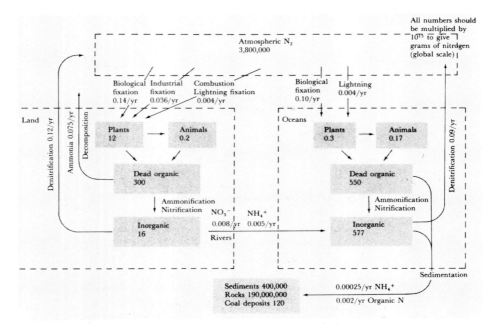

FIGURE 34.   The global nitrogen cycle. Minor compartments and fluxes are not given. (From Brock, T. D., *Biology of Microorganisms*, Prentice-Hall, Englewood Cliffs, N.J., 1979, 802. With permission.)

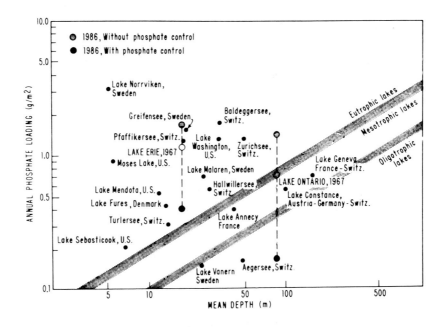

FIGURE 35.   Critical phosphorus loading on 20 lakes. The demarcation line indicates the most relevant collective reference values on permissive phosphorus loadings. From Vollenweider, R. A., *Mem. Ist. Ital. Idrobiol.*, 33, 53, 1976. With permission.)

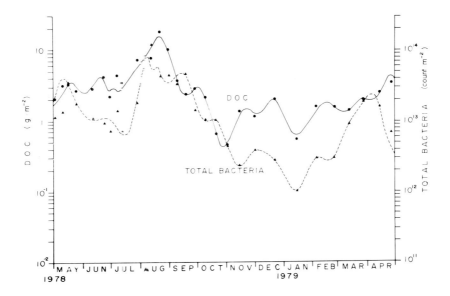

FIGURE 36. Seasonal fluctuation of dissolved organic carbon and bacterioplankton in the water column of the water chestnut ecosystem in Lake Kasumigaura, Japan. (From Nakano, H. and Seki, H., *Water Air Soil Pollut.*, 15, 215, 1981. With permission.)

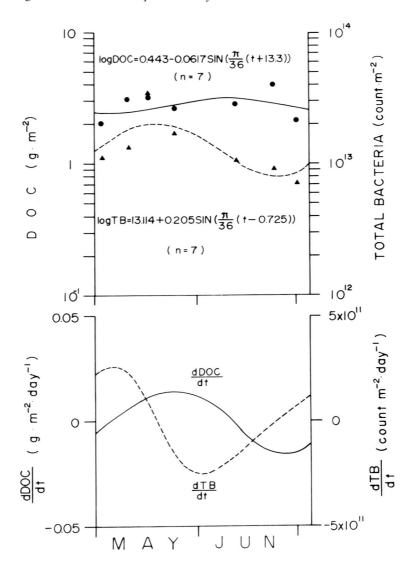

FIGURE 37.   Mathematical models on the fluctuations of dissolved organic carbon and bacterioplankton after the spring bloom of phytoplankton in the water chestnut ecosystem in Lake Kasumigaura, Japan. (From Nakano, H. and Seki, H., *Water Air Soil Pollut.*, 15, 215, 1981. With permission.)

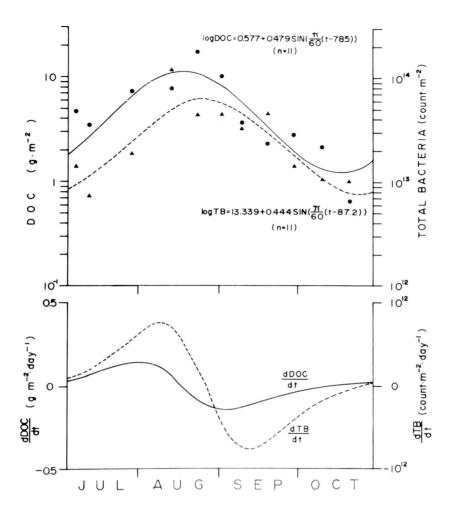

FIGURE 38. Mathematical models on the fluctuations of dissolved organic carbon and bacterioplankton during the blue-green algal bloom in summer in the water chestnut ecosystem of Lake Kasumigaura, Japan. (From Nakano, H. and Seki, H., *Water Air Soil Pollut.*, 15, 215, 1981. With permission.)

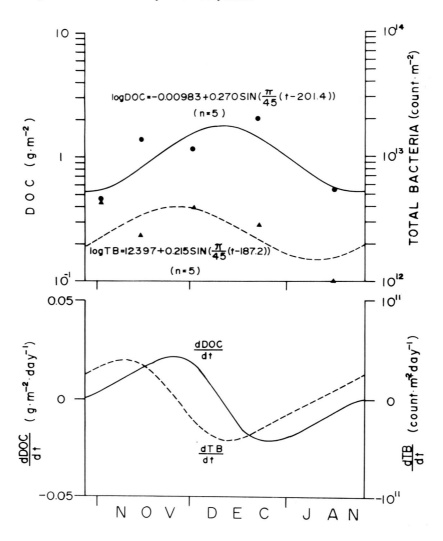

FIGURE 39.   Mathematical models on the fluctuations of dissolved organic carbon and bacterioplankton after the blue-green algal bloom in the water chestnut ecosystem in Lake Kasumigaura, Japan. (From Nakano, H. and Seki, H., *Water Air Soil Pollut.*, 15, 215, 1981. With permission.)

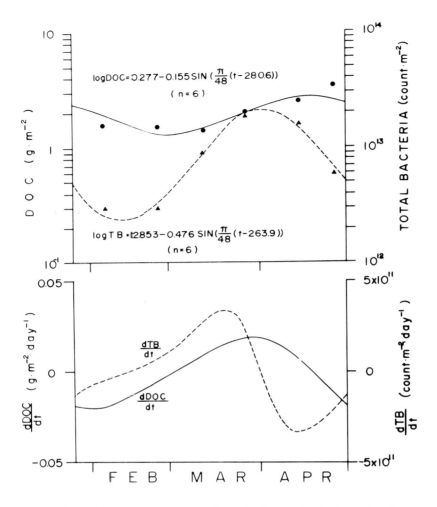

FIGURE 40. Mathematical models on the fluctuations of dissolved organic carbon and bacterioplankton during the eucaryotic algal bloom in spring in the water chestnut ecosystem of Lake Kasumigaura, Japan. (From Nakano, H. and Seki, H., *Water Air Soil Pollut.*, 5, 215, 1981. With permission.)

Table 1
TURNOVER TIME OF CHITIN IN LAKE WATER OF THE
WATER CHESTNUT ECOSYTEM

Date	August 22	August 31	September 8	September 18	September 28	October 9
t (days)	0	9	17	27	37	48
Y (g^2 m)	2.830	0.799	0.463	0.292	0.240	0.217
d Y/d t	0.540	0.080	0.024	0.008	0.003	0.001
T.T.	5.42	9.99	18.17	36.50	80.00	217.00

Note: $Y = e^{2.65\,e^{-0.072t}} - 1.65$, d Y/d $t = -0.072 \times 2.65\,e^{-0.072t} \times e\,(2.65\,e^{-0.072t} - 1.61)$, and T.T. = Turnover Time

From Yamamoto, H. and Seki, H., *Water Air Soil Pollut.*, 12, 519, 1979. With permission.

Table 2
FATTY ACID COMPOSITION OF ORGANISMS IN
EACH TROPHIC LEVEL OF A FOOD CHAIN
EXPERIMENTALLY FORMED IN AN AQUARIUM

			Guppy	
Fatty acid	*Chaetoceros*	*Artemia*	$17 \pm 1°C$	$24 \pm 1.5°C$
Shorter chain	trace	0.4	trace	
12:0	0.4	trace	0.2	trace
13:0	0.7	trace	trace	trace
14:0	12.0	4.8	1.5	0.9
15:0	1.8	1.5	trace	0.2
14:2	0.6	trace	0.6	0.5
16:0	18.1	11.6	22.9	36.0
16:1	47.9	44.9	15.9	8.9
16:2	2.7	trace	0.2	0.2
16:3?	4.0	1.7	—	0.6
16:4?	trace	—	—	0.5
18:0	0.5	1.9	8.2	9.8
18:1	8.7	18.4	18.3	15.0
18:2	1.7	0.7	trace	trace
18:3	trace	0.5	1.4	0.8
20:1	—	0.9	—	—
18:4 & 20:2	—	0.8	0.3	trace
20:3	—	—	0.2	trace
20:4	—	trace	2.0	2.0
20:5	—	12.0	4.8	4.6
22:4	—	—	1.3	1.0
22:5	—	—	6.1	7.3
22:6	—	—	16.5	11.5

From Kayama, M., *Bull. Jpn. Soc. Sci. Fish.*, 30, 647, 1964. With permission.

Table 3

UPTAKE KINETICS OF ORGANIC SUBSTRATES BY
MICROORGANISMS IN THE CENTER OF EACH
WATER MASS OF THE PACIFIC OCEAN[160,161]

Substrate	V (mg/m³/hr)	Kt + Sn (mg/m³)	Tt (hr)	Mineralization / gross assimilation (%)
Station 9 in the western North Pacific central water (50 m depth): January 18, 1973				
Aspartic acid	4.65	38	8,200	28
Glutamic acid	7.62	18	2,400	27
Glysine	4.07	37	9,100	29
Lysine	2.85	13	4,600	16
Glucose	10.8	63	5,900	51
Galactose	13.4	365	27,000	41
Protein hydrolysate	88.7	351	4,000	27
Station 11 in the western North Pacific central water (50 m depth): June 26, 1971				
Aspartic acid	4.57	47	10,000	26
Glutamic acid	13.2	94	7,100	33
Glycine	9.98	182	18,000	39
Alanine	2.63	8.7	3,300	15
Lysine	3.86	23	5,900	14
Glucose	14.1	84	6,000	71
Galactose	10.3	146	14,000	56
Protein hydrolysate	47.9	429	9,000	25
Station 12 in the Kuroshio Current (50 m depth): July 3, 1971				
Aspartic acid	6.95	42	6,000	26
Glutamic acid	12.6	23	1,800	24
Glycine	11.1	21	1,900	27
Alanine	10.6	32	3,000	21
Lysine	3.68	14	3,800	17
Glucose	8.22	22	2,700	42
Galactose	10.4	32	3,100	33
Protein hydrolysate	118	360	3,100	31
Station 19 in the subarctic Pacific water (20 m depth): July 21, 1971				
Aspartic acid	13.5	87	6,500	19
Glutamic acid	24.9	69	2,800	25
Glycine	14.9	38	2,600	29
Alanine	16.5	29	1,800	17
Lysine	8.20	22	2,600	9.8
Glucose	23.2	30	1,300	38
Galactose	19.2	43	2,200	30
Protein hydrolysate	170	500	3,000	11

Note: Kt: transport constant, Sn: *in situ* substrate concentration, Tt: time requirement for complete removal of natural substrate (Sn), and V: maximum attainable rate of uptake.

Table 4

UPTAKE KINETICS OF ORGANIC SUBSTRATES BY
MICROORGANISMS IN THE BOTTOM OF SURFACE
LAYER (100 m) OF THE WESTERN NORTH PACIFIC
CENTRAL WATER[a] (STATION 11)

Substrate	V (mg/m³/hr)	Kt + Sn (mg/m³)	Tt (hr)	Mineralization / gross assimilation (%)
Aspartic acid	5.27	42	8,000	27
Glutamic acid	10.3	98	9,600	33
Glycine	12.7	114	9,000	47
Alanine	2.93	17	5,800	15
Lysine	1.99	9.8	4,900	19
Glucose	4.06	31	7,600	46
Galactose	9.35	224	24,000	47
Protein hydrolysate	32.3	1,140	35,000	11

Note: Kt: transport constant, Sn: *in situ* substrate concentration, Tt: time
requirement for complete removal of natural substrate (Sn), and V:
maximum attainable rate of uptake.

[a]   July 27, 1971.

From Seki, H., Nakai, T., and Otobe, H., *Arch. Hydrobiol.*, 71, 79, 1972. With
permission.

Table 5
UPTAKE KINETICS OF ORGANIC SUBSTRATES BY MICROORGANISMS IN TOKYO BAY ON JULY 9, 1974

Substrate	V (mg/m³/hr)	Kt + Sn (mg/m³)	Tt (hr)	Sn (mg/m³)	Vt (mg/m³/hr)	Mineralization/ gross assimilation (%)
Surface layer (0 m) of Station T-1 in Tokyo Bay						
Aspartic acid	3.0	57	19	44	2.4	34
Glutamic acid	8.1	39	4.8	22	4.6	24
Glycine	9.8	117	12	110	9.2	46
Alanine	6.4	72	11	50	4.5	16
Lysine	3.2	50	16	25	1.6	25
Glucose	18	156	8.7	116	13	79
Galactose	5.7	53	9.2	21	2.3	81
Protein hydrolysate	26	348	13	224	17	19
Intermediate layer (10 m) of Station T-1 in Tokyo Bay						
Aspartic acid	0.6	15	25	4	0.16	26
Glutamic acid	3.0	82	27	62	2.3	37
Glycine	2.3	32	14	20	1.4	47
Alanine	2.6	58	22	46	2.1	18
Lysine	1.0	31	30	18	0.60	21
Glucose	2.2	51	23	29	1.3	43
Galactose	0.52	12	23	3.5	0.15	54
Protein hydrolysate	10.8	314	29	190	6.6	25
Bottom layer (24 m) of Station T-1 in Tokyo Bay						
Aspartic acid	10.4	24	23	14	6.1	24
Glutamic acid	7.7	115	15	92	6.1	35
Glycine	5.0	70	14	59	4.2	43
Alanine	3.4	62	18	54	3.0	19
Lysine	1.8	38	21	28	1.3	21
Glucose	4.6	91	20	24	1.2	56
Galactose	0.93	56	60	22	0.37	51
Protein hydrolysate	21	274	13	158	12	23
Surface layer (0 m) of Station T-2 in Tokyo Bay						
Aspartic acid	1.0	24	24	8	0.33	28
Glutamic acid	5.8	70	15	36	3.0	18
Glycine	3.5	46	13	36	2.8	26
Alanine	3.1	37	12	24	2.0	17
Lysine	1.8	32	18	12	0.67	19
Glucose	4.5	49	11	12	1.1	89
Galactose	1.0	29	28	3	0.11	33
Protein hydrolysate	22.4	291	13	183	14	20

Note: V: maximum attainable rate of uptake; Kt: transport constant; Sn: *in situ* substrate concentration; Tt: time requirement for complete removal of natural substrate (Sn); Vt: *in situ* rate of uptake.

From Seki, H., Yamaguchi, Y., and Ichimura, S., *Arch. Hydrobiol.,* 75, 297, 1975. With permission.

Table 6

UPTAKE KINETICS OF ORGANIC SUBSTRATES BY MICROORGANISMS IN THE SURFACE LAYER (0 m) OF STATION S-1 IN THE INOHZAWA RIVER [162,164]

Substrate	V (mg/m³/hr)	Kt + Sn (mg/m³)	Tt (hr)	Sn (mg/m³)	Vt (mg/m³/hr)	Mineralization/ gross assimilation (%)
			February 20, 1976			
Aspartic acid	0.15	16	110	5.0	0.045	29
Glutamic acid	0.15	8.2	54	2.1	0.039	34
Glycine	0.13	24	190	20	0.11	42
Alanine	0.13	12	95	9.3	0.098	27
Lysine	0.39	18	46	4.1	0.089	46
Glucose	0.31	34	110	6.4	0.058	13
Galactose	0.054	15	280	6.2	0.022	29
			June 28, 1976			
Aspartic acid	0.63	150	240	31	0.13	24
Glutamic acid	0.20	24	120	12	0.10	31
Glycine	0.094	16	170	6.2	0.036	52
Alanine	0.062	5.2	84	1.2	0.014	18
Lysine	0.17	17	100	11	0.11	14
Glucose	0.11	28	250	11	0.044	32
Galactose	0.28	18	65	4.8	0.074	52
			August 26, 1976			
Aspartic acid	1.8	55	30	29	0.97	34
Glutamic acid	2.7	80	30	9.5	0.32	34
Glycine	4.0	85	21	50	2.4	46
Alanine	3.9	120	31	85	2.7	21
Lysine	1.1	29	27	5.5	0.20	12
Glucose	2.1	59	28	8.9	0.32	30
Galactose	4.0	140	35	66	1.9	19
			September 18, 1974			
Aspartic acid	2.6	127	48	8	0.17	22
Glutamic acid	3.8	84	22	14	0.64	44
Glycine	0.69	11	16	2	0.19	33
Alanine	1.5	35	23	3	0.13	23
Lysine	1.7	167	96	44	0.46	17
Glucose	10	330	32	40	1.3	61
Galactose	0.32	26	82	1.2	0.015	28
Protein hydrolysate	20	800	40	180	4.4	15

Note: V: maximum attainable rate of uptake; Kt: transport constant; Sn: *in situ* substrate concentration; Tt: time requirement for complete removal of natural substrate (Sn); Vt: *in situ* rate of uptake.

Table 7
UPTAKE KINETICS OF ORGANIC SUBSTRATES BY MICROORGANISMS IN THE SURFACE LAYER (0 m) OF STATION S-2 IN SHIMODA BAY[162,164]

Substrate	V (mg/m²/hr)	Kt + Sn (mg/m³)	Tt (hr)	Sn (mg/m³)	Vt (mg/m³/hr)	Mineralization /gross assimilation (%)
			February 20, 1976			
Aspartic acid	0.28	19	68	8.5	0.13	55
Glutamic acid	0.74	23	31	12	0.39	74
Glycine	0.095	18	190	1.6	0.0084	33
Alanine	0.15	1.4	92	7.1	0.077	28
Lysine	0.062	18	290	14	0.048	45
Glucose	0.37	99	270	61	0.23	55
Galactose	0.023	27	1200	9.3	0.0078	58
			June 28, 1976			
Aspartic acid	0.33	13	40	5.4	0.14	36
Glutamic acid	0.67	29	43	10	0.23	45
Glycine	0.29	12	42	7.2	0.17	24
Alanine	0.90	19	21	13	0.62	24
Lysine	0.25	55	220	15	0.068	14
Glucose	1.3	46	36	10	0.28	27
Galactose	0.19	21	110	7.8	0.071	41
			August 26, 1976			
Aspartic acid	4.4	140	32	18	0.56	55
Glutamic acid	4.4	120	27	15	0.56	57
Glycine	1.7	53	32	12	0.38	43
Alanine	2.6	76	29	19	0.66	48
Lysine	1.6	61	38	33	0.87	34
Glucose	3.4	75	22	17	0.77	38
Galactose	5.3	170	32	28	0.88	31
			September 18, 1974			
Aspartic acid	3.5	127	36	28	0.78	19
Glutamic acid	11	128	12	37	3.1	34
Glycine	6.0	108	18	7	0.39	28
Alanine	4.9	54	11	23	2.1	21
Lysine	5.0	116	23	18	0.78	17
Glucose	23	710	31	209	6.7	58
Galactose	0.48	24	50	1.2	6.7	58
Protein hydrolysate	20	560	28	162	5.8	11

Note: V: maximum attainable rate of uptake; Kt: transport constant; Sn: in situ substrate concentration; Tt: time requirement for complete removal of natural substrate (Sn); Vt: in situ rate of uptake.

Table 8
UPTAKE KINETICS OF ORGANIC SUBSTRATES BY MICROORGANISMS IN THE INTERMEDIATE LAYER (1 m) OF STATION S-2 IN SHIMODA BAY

Substrate	V (mg/m³/hr)	Kt + Sn (mg/m³)	Tt (hr)	Sn (mg/m³)	Vt (mg/m³/hr)	Mineralization /gross assimilation (%)
			February 20, 1976			
Aspartic acid	2.5	68	27	46	1.7	24
Glutamic acid	4.1	73	18	25	1.4	37
Glycine	1.1	16	15	5.8	0.39	18
Alanine	4.3	56	13	21	1.6	27
Lysine	1.4	12	8.4	4.7	0.56	40
Glucose	3.9	110	28	21	0.75	13
Galactose	0.31	34	110	25	0.23	27
			June 28, 1976			
Aspartic acid	2.6	73	28	16	0.57	33
Glutamic acid	5.8	110	19	21	1.1	42
Glycine	1.6	28	18	14	0.78	22
Alanine	2.8	55	20	13	0.65	20
Lysine	1.3	23	18	8.5	0.47	12
Glucose	4.2	96	23	34	1.5	19
Galactose	1.8	61	34	16	0.47	27
			August 26, 1976			
Aspartic acid	1.2	45	39	20	0.51	66
Glutamic acid	3.0	120	40	15	0.38	43
Glycine	1.5	66	44	29	0.66	32
Alanine	2.3	78	34	31	0.91	49
Lysine	4.2	130	31	70	2.3	41
Glucose	5.0	160	32	64	2.0	25
Galactose	0.24	85	36	45	1.3	42

Note: V: maximum attainable rate of uptake; Kt: transport constant; Sn: *in situ* substrate concentration; Tt: time requirement for complete removal of natural substrate (Sn); Vt: *in situ* rate of uptake.

From Seki, H., Terada, T., and Ichimura, S., *Arch. Hydrobiol.*, 88, 219, 1980. With permission.

Table 9
UPTAKE KINETICS OF ORGANIC SUBSTRATES BY MICROORGANISMS IN THE BOTTOM LAYER (2 m) OF STATION S-2 IN SHIMODA BAY

Substrate	V (mg/m³/hr)	Kt + Sn (mg/m³)	Tt (hr)	Sn (mg/m³)	Vt (mg/m³/hr)	Mineralization /gross assimilation (%)
February 20, 1976						
Aspartic acid	0.83	20	24	8.1	0.34	39
Glutamic acid	1.6	31	19	11	0.58	26
Glycine	1.2	13	11	5.2	0.47	35
Alanine	2.2	52	24	34	1.4	11
Lysine	1.7	32	19	28	1.5	59
Glucose	2.6	51	20	9.9	0.50	24
Galactose	0.29	15	52	7.4	0.14	26
June 28, 1976						
Aspartic acid	6.0	150	25	14	0.56	45
Glutamic acid	3.8	83	22	4.2	0.19	36
Glycine	2.0	47	23	12	0.52	13
Alanine	3.0	68	23	13	0.57	19
Lysine	1.4	27	20	5.7	0.29	8.5
Glucose	4.0	92	23	10	0.43	16
Galactose	2.3	67	29	26	0.90	38
August 26, 1976						
Aspartic acid	2.3	68	29	20	0.69	61
Glutamic acid	3.0	130	43	64	1.5	56
Glycine	1.9	66	34	21	0.62	34
Alanine	4.3	150	35	95	2.7	55
Lysine	2.3	48	21	36	1.7	35
Glucose	3.8	130	34	33	0.97	33
Galactose	2.5	110	44	74	1.7	25

Note: V: maximum attainable rate of uptake; Kt: transport constant; Sn: *in situ* substrate concentration; Tt: time requirement for complete removal of natural substrate (Sn); Vt: *in situ* rate of uptake.

From Seki, H., Terada, T., and Ichimura, S., *Arch. Hydrobiol.*, 88, 219, 1980. With permission.

Table 10
UPTAKE KINETICS OF ORGANIC SUBSTRATES BY MICROORGANISMS IN THE SURFACE LAYER (0 m) OF STATION S-3 IN THE KUROSHIO COUNTER CURRENT[162,164]

Substrate	V (mg/m³/hr)	Kt + Sn (mg/m³)	Tt (hr)	Sn (mg/m³)	Vt (mg/m³/hr)	Mineralization /gross assimilation (%)
			February 20, 1976			
Asparatic acid	0.0023	9.1	3900	4.0	0.0010	17
Glutamic acid	0.014	12	850	4.6	0.0054	15
Glycine	0.0066	7.9	1200	2.7	0.0023	24
Alanine	0.018	38	2100	29	0.014	13
Lysine	0.022	62	2800	41	0.015	23
Glucose	0.024	19	790	5.1	0.0065	39
Galactose	0.0024	15	6200	12	0.0019	44
			June 28, 1976			
Aspartic acid	0.18	6.9	38	6.1	0.16	52
Glutamic acid	0.25	16	65	4.3	0.066	19
Glycine	0.13	5.2	40	1.3	0.033	36
Alanine	0.12	6.2	51	4.3	0.084	25
Lysine	0.11	6.1	56	2.9	0.052	17
Glucose	0.20	18	88	7.1	0.081	20
Galactose	0.13	19	150	17	0.11	26
			August 26, 1976			
Aspartic acid	2.6	73	28	46	1.6	37
Glutamic acid	0.43	32	75	18	0.24	58
Glycine	2.5	110	43	31	0.72	51
Alanine	2.6	63	24	8.4	0.35	31
Lysine	0.68	25	37	6.4	0.17	40
Glucose	2.0	92	46	27	0.59	31
Galactose	0.37	30	81	26	0.32	26
			September 18, 1974			
Aspartic acid	1.1	118	110	8.6	0.078	20
Glutamic acid	0.61	115	190	12	0.061	27
Glycine	0.38	137	360	6.1	0.017	24
Alanine	0.19	37	190	14	0.073	25
Lysine	0.13	43	320	11	0.034	20
Glucose	0.30	57	190	28	0.15	52
Galactose	0.14	43	310	1.5	0.0048	27
Protein hydrolysate	1.0	120	120	89	0.74	17

Note: V: maximum attainable rate of uptake; Kt: transport constant; Sn: *in situ* substrate concentration; Tt: time requirement for complete removal of natural substrate (Sn); Vt: *in situ* rate of uptake.

Table 11

UPTAKE KINETICS OF ORGANIC SUBSTRATES BY MICROORGANISMS IN THE SURFACE LAYER (0 m) OF STATION S-4 IN A BRANCH OF THE KUROSHIO CURRENT; SEPTEMBER 18, 1974

Substrate	V (mg/m³/hr)	Kt + Sn (mg/m³)	Tt (hr)	Sn (mg/m³)	Vt (mg/m³/hr)	Mineralization /gross assimilation (%)
Aspartic acid	0.22	99	460	4.0	0.0086	26
Glutamic acid	0.75	112	150	5.1	0.034	26
Glycine	0.57	159	280	1.3	0.0046	27
Alanine	0.46	73	160	3.1	0.019	24
Lysine	0.28	93	330	3.5	0.011	14
Glucose	0.071	36	510	11	0.022	71
Galactose	0.024	41	1700	1.1	0.00065	27
Protein hydrolysate	3.5	518	150	30	0.20	17

Note: V: maximum attainable rate of uptake; Kt: transport constant; Sn: *in situ* substrate concentration; Tt: time requirement for complete removal of natural substrate (Sn); Vt: *in situ* rate of uptake.

From Seki, H., Yamaguchi, Y., and Ichimura, S., *Arch. Hydrobiol.,* 75, 297, 1975. With permission.

Table 12

UPTAKE KINETICS[a] OF ORGANIC SUBSTRATES BY MICROORGANISMS IN THE SURFACE LAYER (0 m) OF THE STAMP RIVER, B.C., CANADA

Substrate	V (mg/m³/hr)	Kt + Sn (mg/m³)	Tt (hr)	Sn (mg/m³)	Vt (mg/m³/hr)	Mineralization /gross assimilation (%)
Aspartic acid	0.28	31	110	17	0.15	25
Glutamic acid	0.35	34	97	14	0.14	21
Glycine	0.18	22	120	14	0.12	29
Alanine	0.54	76	140	31	0.22	21
Lysine	0.17	64	380	18	0.047	32
Glucose	0.32	35	110	22	0.020	36
Galactose	0.21	176	360	12	0.033	38
Acetic acid	0.27	32	120	4.5	0.038	51
Glycolic acid	0.24	15	62	12	0.19	40

Note: V: maximum attainable rate of uptake; Kt: transport constant; Sn: *in situ* substrate concentration; Tt: time requirement for complete removal of natural substrate (Sn); Vt: *in situ* rate of uptake.

[a] October 5, 1978.

From Seki, H., MacIsaac, E. A., and Stockner, J. G., *Arch Hydrobiol.,* 88, 58, 1980. With permission.

Table 13
UPTAKE KINETICS OF ORGANIC SUBSTRATES BY MICROORGANISMS IN THE SURFACE LAYER OF THE UNFERTILIZED REGION OF GREAT CENTRAL LAKE, B.C., CANADA[165,174]

Substrate	V (mg/m³/hr)	Kt + Sn (mg/m³)	Tt (hr)	Sn (mg/m³)	Vt (mg/m³/hr)	Mineralization /gross assimilation (%)
			July 24, 1979			
Aspartic acid	0.20	46	230	11	0.048	25
Glutamic acid	0.53	64	120	12	0.10	39
Glycine	0.19	52	280	16	0.057	38
Alanine	0.27	56	210	14	0.067	23
Lysine	0.24	44	180	12	0.067	30
Glucose	0.30	74	250	19	0.076	29
Galactose	0.067	87	1,300	9	0.0069	29
Acetic acid	0.15	31	210	18	0.086	42
Glycolic acid	0.29	61	210	32	0.15	34
			October 7, 1978			
Aspartic acid	0.21	45	210	22	0.10	25
Glutamic acid	0.33	39	120	10	0.083	23
Glycine	0.24	46	190	18	0.095	25
Alanine	0.27	84	310	18	0.058	33
Lysine	0.15	55	360	18	0.050	24
Glucose	0.22	38	170	8.5	0.050	29
Galactose	0.21	110	530	22	0.042	36
Acetic acid	0.51	47	93	16	0.17	38
Glycolic acid	0.10	9.1	87	4.8	0.055	35

Note: V: maximum attainable rate of uptake; Kt: transport constant; Sn: *in situ* substrate concentration; Tt: time requirement for complete removal of natural substrate (Sn); Vt: *in situ* rate of uptake.

163

Table 14
UPTAKE KINETICS OF ORGANIC SUBSTRATES BY MICROORGANISMS IN THE SURFACE LAYER OF THE FERTILIZED REGION OF GREAT CENTRAL LAKE, B.C., CANADA[165,174]

Substrate	V (mg/m³/hr)	Kt + Sn (mg/m³)	Tt (hr)	Sn (mg/m³)	Vt (mg/m³/hr)	Mineralization /gross assimilation (%)
			July 24, 1979			
Aspartic acid	0.48	67	140	19	0.13	29
Glutamic acid	0.35	52	150	24	0.16	33
Glycine	0.22	35	160	19	0.12	35
Alanine	0.24	36	150	23	0.15	45
Lysine	0.35	60	170	32	0.19	45
Glucose	0.48	72	150	25	0.17	27
Galactose	0.22	115	530	9	0.017	34
Acetic acid	0.28	54	190	23	0.12	35
Glycolic acid	0.23	39	170	21	0.12	31
			October 7, 1978			
Aspartic acid	0.38	45	120	16	0.13	28
Glutamic acid	0.26	21	81	7.5	0.093	20
Glycine	0.16	28	180	8.8	0.049	36
Alanine	0.16	74	470	32	0.068	26
Lysine	0.28	64	230	36	0.16	28
Glucose	2.2	100	45	18	0.40	31
Galactose	0.17	96	560	12	0.021	36
Acetic acid	1.3	93	70	30	0.43	39
Glycolic acid	1.9	95	49	36	0.73	42

Note: V: maximum attainable rate of uptake; Kt: transport constant; Sn: *in situ* substrate concentration; Tt: time requirement for complete removal of natural substrate (Sn); Vt: *in situ* rate of uptake.

Table 15
UPTAKE KINETICS OF ORGANIC SUBSTANCES BY MICROORGANISMS IN THE SURFACE LAYER OF GLACIALLY OLIGOTROPHIC LAKES IN B.C., CANADA

Substrate	V (mg/m³/hr)	Kt + Sn (mg/m³)	Tt (hr)	Sn (mg/m³)	Vt (mg/m³/hr)	Mineralization /gross assimilation (%)
Meziadin Lake July 10, 1979						
Aspartic acid	0.034	68	2,000	19	0.0095	36
Glutamic acid	0.045	68	1,500	9	0.0060	29
Glycine	0.0067	31	4,600	5	0.0011	31
Alanine	0.018	95	5,300	7.5	0.0014	37
Lysine	0.0059	32	5,400	5.5	0.0010	21
Glucose	0.0062	32	5,200	2	0.0003	23
Galactose	0.0082	41	5,000	1.5	0.0004	39
Acetic acid	0.033	27	810	11	0.014	37
Glycolic acid	0.037	21	570	15	0.026	45
Kitlope Lake July 13, 1979						
Aspartic acid	0.070	48	690	8.5	0.012	34
Glutamic acid	0.074	29	390	3	0.0077	32
Glycine	0.0052	14.5	2,800	6	0.0021	29
Alanine	0.031	44	1,400	3.5	0.0025	38
Lysine	0.14	52	380	3.5	0.0025	38
Glucose	0.024	44	1,800	12	0.0067	33
Galactose	0.0055	17	3,100	1.5	0.00048	43
Acetic acid	0.28	78	280	59	0.21	53
Glycolic acid	0.33	120	360	39	0.11	47

Note: V: maximum attainable rate of uptake; Kt: transport constant; Sn: *in situ* substrate concentration; Tt: time requirement for complete removal of natural substrate (Sn); Vt: *in situ* rate of uptake.

From Seki, H., Shortreed, K. S., and Stockner, J. G., *Arch. Hydrobiol.*, 90, 210, 1980. With permission.

Table 16
UPTAKE KINETICS OF ORGANIC SUBSTANCES BY MICROORGANISMS IN
THE SURFACE LAYER OF DYSTROPHIC LAKES IN B.C., CANADA

Substrate	V (mg/m³/hr)	Kt + Sn (mg/m³)	Tt (hr)	Sn (mg/m³)	Vt (mg/m³/hr)	Mineralization /gross assimilation (%)
			Lowe Lake	July 12, 1979		
Aspartic acid	0.063	34	540	16	0.030	37
Glutamic acid	0.080	55	690	9	0.013	39
Glycine	0.056	22	390	5	0.013	32
Alanine	0.066	47	710	5.5	0.0077	26
Lysine	0.048	32	670	11	0.016	43
Glucose	0.027	27	1,000	6.5	0.0065	49
Galactose	0.008	16	2,000	1	0.0005	34
Acetic acid	0.042	21	500	7	0.014	42
Glycolic acid	0.053	85	1,600	57	0.036	46
			Bonilla Lake	July 14, 1979		
Aspartic acid	0.058	11	190	5	0.036	29
Glutamic acid	0.16	33	210	8	0.038	29
Glycine	0.066	49	740	7	0.0095	39
Alanine	0.058	57	990	5	0.0086	37
Lysine	0.12	31	250	8	0.032	36
Glucose	0.11	10	150	4.5	0.03	25
Galactose	0.032	7	220	1	0.0045	30
Acetic acid	0.023	14	610	6	0.0098	44
Glycolic acid	0.022	28	1,300	13	0.01	34

Note: V: maximum attainable rate of uptake; Kt: transport constant; Sn: *in situ* substrate concentration; Tt: time requirement for complete removal of natural substrate (Sn); Vt: *in situ* rate of uptake.

From Seki, H., Shortreed, K. S., and Stockner, T. G., *Arch. Hydrobiol.*, 90, 210, 1980. With permission.

Table 17

UPTAKE KINETICS OF ORGANIC SUBSTRATES BY MICROORGANISMS IN THE SURFACE LAYER IN DIFFERENT LOCATIONS OF ALBERNI INLET, B.C., CANADA

Substrate	V (mg/m³/hr)	Kt + Sn (mg/m³)	Tt (hr)	Sn (mg/m³)	Vt (mg/m³/hr)	Mineralization/ gross assimilation (%)
Surface layer (1 m) in the entrance to Alberni Inlet October 8, 1978						
Aspartic acid	0.15	22	150	10	0.067	34
Glutamic acid	0.25	64	260	21	0.081	47
Glycine	0.17	38	230	13	0.057	52
Alanine	0.18	95	530	12	0.023	36
Lysine	0.42	75	180	24	0.13	43
Glucose	0.27	64	240	13	0.054	33
Galactose	0.22	176	340	7.3	0.021	40
Acetic acid	0.35	99	280	78	0.28	43
Glycolic acid	0.11	59	560	11	0.020	48
Surface layer (1 m) in the central region of Alberni Inlet October 6, 1978						
Aspartic acid	2.2	39	18	25	1.4	28
Glutamic acid	3.3	36	11	13	1.2	38
Glycine	0.44	48	110	24	0.22	27
Alanine	0.36	40	110	28	0.25	21
Lysine	3.4	110	32	28	0.88	47
Glucose	2.6	18	70	11	1.6	45
Galactose	0.93	26	28	8.3	0.30	27
Acetic acid	1.9	29	15	12	0.80	33
Glycolic acid	0.37	36	9.8	18	0.18	36
Surface layer (1 m) in the head of Alberni Inlet October 6, 1978						
Aspartic acid	0.23	8.4	36	4.8	0.13	25
Glutamic acid	0.094	13	32	1.6	0.05	18
Glycine	0.27	11	41	5.3	0.13	25
Alanine	0.17	6.4	38	3.4	0.089	17
Lysine	0.34	20	58	7.6	0.13	19
Glucose	2.2	42	19	22	1.2	29
Galactose	0.36	17	47	15	0.32	20
Acetic acid	0.84	69	82	39	0.48	39
Glycolic acid	0.18	27	150	10	0.067	48

Note: V: maximum attainable rate of uptake; Kt: transport constant; Sn: *in situ* substrate concentration; Tt: time requirement for complete removal of natural substrate (Sn); Vt: *in situ* rate of uptake.

From Seki, H., MacIsaac, E. A., and Stockner, J. G., *Arch. Hydrobiol.*, 88, 58, 1980. With permission.

Table 18

UPTAKE KINETICS OF ORGANIC SUBSTANCES BY MICROORGANISMS ON AUGUST 6, 1980, IN THE SURFACE LAYER (0 m) OF PATRICIA BAY, SAANICH INLET, B.C., CANADA[183]

Substrate	V (mg/m³/hr)	Kt + Sn (mg/m³)	Tt (hr)	Sn (mg/m³)	Vt (mg/m³/hr)	Mineralization/ gross assimilation (%)
Aspartic acid	0.069	11	160	4.0	0.025	42
Glutamic acid	0.18	50	275	22	0.080	69
Glycine	0.099	33	334	26	0.078	44
Alanine	0.13	21	166	17	0.10	21
Lysine	0.075	14	186	5.6	0.030	5.4
Glucose	0.061	5.0	82	2.2	0.027	19
Galactose	0.17	61	358	46	0.13	55
Acetic acid	0.027	17	624	11	0.018	18
Glycolic acid	0.011	6.2	580	4.8	0.0083	78

Note: V: maximum attainable rate of uptake; Kt: transport constant, Sn: *in situ* substrate concentration; Tt: time requirement for complete removal of natural substrate (Sn); Vt: *in situ* rate of uptake.

Table 19

UPTAKE KINETICS OF ORGANIC SUBSTRATES BY MICROORGANISMS IN THE BOTTOM LAYER (APPROXIMATELY 20 m) OF PATRICIA BAY, SAANICH INLET, B.C.,CANADA

Substrate	V (mg/m³/hr)	Kt + Sn (mg/m³)	Tt (hr)	Sn (mg/m³)	Vt (mg/m³/hr)	Mineralization/ gross assimilation (%)
			9 a.m. July 18, 1980			
Aspartic acid	0.072	11	152	5.6	0.037	45
Glutamic acid	0.077	4.3	56	2.2	0.039	20
Glycine	0.072	13	180	8.9	0.049	54
Alanine	0.085	4.4	52	3.5	0.067	22
Lysine	0.0052	4.1	788	2.3	0.0029	30
Glucose	0.0057	2.2	386	0.80	0.0021	4.9
Galactose	0.017	6.7	400	1.5	0.0038	17
Acetic acid	0.039	29	736	22	0.030	13
Glycolic acid	0.0091	7.4	810	3.1	0.0038	52
			9 a.m. July 19, 1980			
Aspartic acid	0.067	11	165	2.1	0.013	21
Glutamic acid	0.15	23	158	12	0.076	21
Glycine	0.030	8.8	293	3.7	0.013	43
Alanine	0.046	17	371	5.0	0.013	38
Lysine	0.038	8.2	218	3.9	0.018	24
Glucose	0.013	3.1	248	1.4	0.0056	11
Galactose	0.012	5.9	502	3.6	0.0072	24
Acetic acid	0.019	18	936	11	0.012	12
Glycolic acid	0.053	28	531	7.2	0.014	33
			9 a.m. July 20, 1980			
Aspartic acid	0.21	16	76	12	0.16	43
Glutamic acid	0.028	8.2	56	1.8	0.0061	22
Glycine	0.13	15	114	12	0.11	16

Table 19 (continued)
UPTAKE KINETICS OF ORGANIC SUBSTRATES BY MICROORGANISMS IN THE BOTTOM LAYER (APPROXIMATELY 20 m) OF PATRICIA BAY, SAANICH INLET, B.C.,CANADA

Substrate	V (mg/m³/hr)	Kt + Sn (mg/m³)	Tt (hr)	Sn (mg/m³)	Vt (mg/m³/hr)	Mineralization/ gross assimilation (%)
Alanine	0.16	10	62	6.2	0.10	10
Lysine	0.053	4.2	80	1.8	0.023	20
Glucose	0.019	5.3	272	3.2	0.012	8.7
Galactose	0.062	14	226	9.3	0.041	9.1
Acetic acid	0.044	25	565	13	0.023	12
Glycolic acid	0.023	19	824	11	0.013	47

9 a.m. July 21, 1980

Substrate	V	Kt + Sn	Tt	Sn	Vt	Mineralization
Aspartic acid	0.089	7.3	82	4.6	0.056	32
Glutamic acid	0.14	7.9	56	3.3	0.059	48
Glycine	0.087	11	126	6.0	0.048	21
Alanine	0.044	3.2	72	1.1	0.015	15
Lysine	0.044	3.1	70	1.2	0.017	26
Glucose	0.17	45	260	18	0.069	8.5
Galactose	0.099	18	182	4.9	0.027	38
Acetic acid	0.066	24	364	3.8	0.010	12
Glycolic acid	0.049	25	510	9.9	0.019	60

9 a.m. July 22, 1980

Substrate	V	Kt + Sn	Tt	Sn	Vt	Mineralization
Aspartic acid	0.067	8.4	126	3.1	0.025	44
Glutamic acid	0.19	7.8	42	2.0	0.048	23
Glycine	0.036	5.0	140	1.2	0.0086	51
Alanine	0.13	15	120	4.1	0.034	42
Lysine	0.076	9.9	130	7.8	0.060	16
Glucose	0.31	30	98	7.0	0.071	8.0
Galactose	0.050	12	240	4.2	0.018	18
Acetic acid	0.038	14	364	11	0.030	19
Glycolic acid	0.022	7.9	364	2.9	0.0080	73

9 a.m. August 18, 1980

Substrate	V	Kt + Sn	Tt	Sn	Vt	Mineralization
Aspartic acid	0.42	57	136	13	0.096	53
Glutamic acid	0.30	24	81	11	0.14	45
Glycine	0.23	34	146	8.9	0.061	53
Alanine	0.23	41	180	11	0.061	34
Lysine	0.26	48	186	23	0.12	26
Glucose	0.13	22	168	6.4	0.038	49
Galactose	0.44	40	90	14	0.16	56
Acetic acid	0.092	42	458	21	0.046	28
Glycolic acid	0.13	65	519	18	0.035	76

Note: V: maximum attainable rate of uptake; Kt; transport constant; Sn: *in situ* substrate concentration; Tt: time requirement for complete removal of natural substrate (Sn); Vt: *in situ* rate of uptake.

From Seki, H., Whitney, F., and Wong, C. S., *Arch. Hydrobiol.,* in press. With permission.

Table 20
TURNOVER TIME OF ORGANIC COMPOUNDS IN DIFFERENT WATER MASSES

	Turnover time		
Type of water mass	Amino acids monosaccharides organic acids, etc.	Cellulose, chitin, etc.	Aquatic humus, etc.
Oligotrophic			
Surface layer	Several tens of days	A few years	Between several tens of years and several hundreds of years[b]
Deep layer	Between a few months and a few years[b]	A few tens of years[b]	Several thousands of years
Mesotrophic	Between a few days and several tens of days	Several months	Several tens of years
Eutrophic	A few days	Between several tens of days and several months	Several years[a]
Hypereutrophic	Less than a few days	Between several days and several tens of days	Between half a year and 1 year

[a] Speculated but not yet determined.
[b] Calculated from data in Ref. 31, 90, 95, 114, 160, 161, 177.

REFERENCES

1. **Miller, S. L.,** Production of some organic compounds under possible primitive earth condition, *J. Am. Chem. Soc.,* 77, 2351, 1955.
2. **Postgate, J. R.,** *The Sulphate-Reducing Bacteria,* Cambridge University Press, London, 1979, 151.
3. **Sorokin, Yu. I.,** On the primary production and bacterial activities in the Black Sea, *J. Cons. Int. Explor. Mer,* 29, 41, 1964.
4. **Sorokin, Yu. I.,** Interrelations between sulphur and carbon turnover in meromictic lakes, *Arch. Hydrobiol.,* 66, 391, 1970.
5. **Seki, H.,** Formation of anoxic zones in seawater, in *Biological Oceanography of the Northern North Pacific Ocean,* Takenouti, A. Y., Ed., Idemitsu Shoten, Tokyo, 1972, 487.
6. **Kriss, A. E.,** *Marine Microbiology (Deep Sea),* (transl.), Oliver & Boyd, London, 1963, 536.
7. **Adair, F. W. and Gundersen, K.,** Chemoautotrophic sulfur bacteria from the marine environment. II. Characterization of an obligately marine facultative autotroph, *Can. J. Microbiol.,* 15, 355, 1969.
8. **Tuttle, J. H., Holmes, P. E., and Jannasch, H. W.,** Growth rate stimulation of marine pseudomonads by thiosulfate, *Arch. Microbiol.,* 99, 1, 1974.
9. **Bernal, J. D.,** The origin of life, *New Biol.,* 16, 1954.
10. **ZoBell, C. E. and Anderson, D. Q.,** Observations on the multiplication of bacteria in different volumes of stored sea water and the influence of oxygen tension and solid surfaces, *Biol. Bull.,* 71, 324, 1936.
11. **Parsons, T. R.,** Some ecological experimental and evolutional aspects of the upwelling ecosystem, *S. Afr. J. Sci.,* 75, 536, 1979.
12. **Oppenheimer, C. H., Ed.,** *Marine Biology IV. Unresolved Problems in Marine Microbiology,* New York Academy of Sciences, New York, 1968, 485.
13. **Parsons, T. R. and Takahashi, M.,** *Biological Oceanographic Progresses,* Pergamon Press, Oxford, 1973, 186.
14. **Stanier, R. Y. and Cohen-Bazire, G.,** The role of light in the microbial world: some facts and speculations, in *Microbial Ecology,* Cambridge University Press, London, 1957, 56.
15. **Mann, K. H.,** Macrophyte production and detritus food chains in coastal waters, *Mem. Ist. Ital. Idrobiol.,* Suppl. 29, 353, 1972.
16. **Seki, H. and Yokohama, Y,** Experimental decay of eelgrass (*Zostera marina*) into detritus particles, *Arch. Hydrobiol.,* 84, 109, 1978
17. **Odum, E. P.,** *Fundamentals of Ecology,* W. B. Saunders, Philadelphia, 1971, 202.
18. **Knox, G. A.,** Antarctic marine ecosystems, in *Antarctic Ecology,* Holdgate, M. W., Ed., Academic Press, London, 1970, 69.
19. **Hutchinson, G. E.,** The biosphere, in *The Biosphere,* W. H. Freeman, San Francisco, 1970, 3.
20. **Murphy, R. C.,** The oceanic life of the Antarctic, *Sci. Am.,* 207, 186, 1962.
21. **Phillipson, J.,** *Ecological Energetics,* Edward Arnold Ltd., London, 1966, 57.
22. **Lindeman, R. L.,** The trophic-dynamic aspect of ecology, *Ecology,* 23, 399, 1942.
23. **Ryther, J. H.,** Photosynthesis and fish production in the sea, *Science,* 166, 72, 1969.
24. **Seki, H.,** The role of microorganisms in the marine food chain with reference to organic aggregate, *Mem. Ist. Ital. Idrobiol.,* Suppl. 29, 245, 1972.
25. **Parsons, T. R. and Strickland, J. D. H.,** On the production of particulate organic carbon by heterotrophic processes in sea water, *Deep Sea Res.,* 8, 211, 1962.
26. **Vinogradov, M. E.,** Food sources of deep-water fauna, speed of decomposition of dead Pteropoda, *Dokl. Acad. Sci. USSR, Biol. Sci Sect.,* 138, 39, 1961.
27. **Kusnezow, S. I., Karsinkin, G. S., Jegorowa, A. A., Kastalskaja, M. A., Karassikowa, A. A., Iwanow, M. W., Sawarsin, G. A., and Derjugina, S. P.,** Schilfgräser als Gründünger zur Erhöhung des Fischertrages der Laichzuchtanstalten, *Wopr. Ichtiol.,* 5, 119, 1955.
28. **Parsons, T. R. and Strickland, J. D. H.,** Oceanic detritus, *Science,* 136, 313, 1962.
29. **Corner, F. D. S. and Coway, C. B.,** Biochemical studies on the production of marine zooplankton, *Biol. Rev.,* 43, 393, 1968.
30. **Degens, E. T.,** Molecular nature of nitrogeneous compounds in sea water and recent marine sediments, in *Organic Matter in Natural Waters,* Hood, D. W., Ed., University of Alaska, Fairbanks, 1970, 77.
31. **Seki, H., Skelding, J., and Parsons, T. R.,** Observations on the decomposition of a marine sediment, *Limnol. Oceanogr.,* 13, 440, 1968
32. **Parsons, T. R. and Seki, H.,** Importance and general implications of organic matter in aquatic environments, in *Organic Matter in Natural Waters,* Hood, D. W., Ed., University of Alaska, Fairbanks, 1970, 1.
33. **Parsons, T. R. and LeBrasseur, R. J.,** The availability of food to different trophic levels in the marine food chain, in *Marine Food Chains,* Steele, J. H., Ed., Oliver & Boyd, Edinburgh, 1970, 325.

34. **Baier, C. R.,** Studien zur Hydrobakteriologie stehender Binnengewässer, *Arch. Hydrobiol.,* 29, 183, 1935.

35. **Seki, H.,** Studies on microbial participation to food cycle in the sea. III. Trial cultivation on brine shrimp to adult in a chemostat (1), *J. Oceanogr. Soc. Jpn.,* 22, 105, 1966.

36. **Sutcliffe, W. H., Jr.,** Some relations of land drainage, nutrients, particulate material and fish catch in two eastern Canadian bays, *J. Fish. Res. Bd. Can.,* 29, 357, 1972.

37. **Nakano, H. and Seki, H.,** Impact of nutrient enrichment in a waterchestnut ecosystem at Takahama-iri Bay of Lake Kasumigaura, Japan V. Dynamics of organic debris, *Water Air Soil Pollut.,* 15, 215, 1981.

38. **Parsons, T. R.,** Suspended organic matter in sea water, in *Progress in Oceanography,* Sears, M., Ed., Pergamon Press, Oxford, 1963, 205.

39. **Seki, H.,** Rôle du micro-organismes dans la chaîne alimentaire de la mer profonde, *La Mer,* 27, 1970.

40. **Seki, H. and ZoBell, C. E.,** Microbial assimilation of carbon dioxide in the Japan Trench, *J. Oceanogr. Soc. Jpn.,* 23, 182, 1967.

41. **Wood, H. G. and Werkman, C. H.,** The utilization of CO_2 by the propionic acid bacteria, *Biochem. J.,* 32, 1262, 1938.

42. **Seki, H.,** Relation between production and mineralization of organic matter in Aburatsubo Inlet, Japan, *J. Fish. Res. Bd. Can.,* 25, 625, 1968.

43. **ZoBell, C. E.,** Microorganisms and the seas, in *Man and the Ocean,* Jpn. Assoc. Int. Ocean Expos., Okinawa, 1975, 502.

44. **Sheldon, R. W., Prakash, A., and Sutcliffe, W. H., Jr.,** The size distribution of particles in the ocean, *Limnol. Oceanogr.,* 17, 327, 1972.

45. **Seki, H.,** The role of microorganisms in the marine productivity, (in Japanese), *Microbial Ecology 3,* University of Tokyo Press, Tokyo, 1976, 73.

46. **Seki, H.,** Microbial biomass on particulate organic matter in seawater of the euphotic zone, *Appl. Microbiol.,* 19, 960, 1970.

47. **Hutchinson, G. E.,** *A Treatise on Limnology,* Volume I, John Wiley & Sons, New York, 1957, 1015.

48. **Piccard, J.,** Nature of the seas, in *Man and the Ocean,* Jpn. Assoc. Int. Ocean Expos., Okinawa, 1975, 361.

49. **ZoBell, C. E.,** Importance of microorganisms in the sea, in *Low Temperature Microbiology Symposium,* Campbell Soup Company, Camden, N.J., 1961, 107.

50. **Sverdrup, H. U., Johnson, M. W., and Fleming, R. H.,** *The Oceans. Their Physics, Chemistry, and General Biology,* Prentice-Hall, Englewood Cliffs, N.J., 1942, 1087.

51. **Morita, R. Y.,** Psychrophilic bacteria, *Bacteriol. Rev.,* 39, 144, 1975.

52. **Ekelöf, E.,** Bakteriologische Studien während der Schwedischen Südpolar Expedition 1901-1903, *W. Erg. der Schwed. Sudpolar Expedition, 1901-1903,* Volume 4, Bonnier, Stockholm, 1908.

53. **Brock, T. D.,** *Biology of Microorganisms,* 3rd ed., Prentice-Hall, Englewood Cliffs, N.J., 1979, 802.

54. **Morita, R. Y.,** Current status of the microbiology of the deep-sea, *Ambio Special Rep.,* 6, 33, 1979.

55. **Sieburth, J. McN.,** Seasonal selection of estuarine bacteria by water temperature, *J. Exp. Mar. Biol. Ecol.,* 1, 98, 1967.

56. **Brock, T. D.,** Life at high temperatures, *Science,* 158, 1012, 1967.

57. **Larsen, H.,** Halophilism, in *The Bacteria. A Treatise on Structure and Function. IV. The Physiology of Growth,* Gunsalus, I. C. and Stanier, R. Y., Eds., Academic Press, New York, 1962, 297.

58. **Seki, H., Stephens, K. V., and Parsons, T. R.,** The contribution of allochthonous bacteria and organic materials from a small river into a semi-enclosed sea, *Arch. Hydrobiol.,* 66, 37, 1969.

59. **Seki, H. and Ebara, A.,** Effect of seawater intrusion on microorganisms in the River Teshio, Hokkaido, Japan, *J. Oceanogr. Soc. Jpn.,* 36, 30, 1980.

60. **Baas Becking, L. G. M. and Wood, E. J. F.,** Biological processes in the estuarine environment. I. Ecology of the sulphur cycle, *K. Ned. Akad. Wet. Versl. Gewone Vergad. Afd. Natuurkd.,* 58, 160, 1955.

61. **Frerman, F. E. and White, D. C.,** Membrane lipid changes during formation of a functional electron transport system, in *Staphylococcus aureus, J. Bacteriol.,* 94, 1868, 1967.

62. **Lewis, W. K. and Whitman, W. G.,** Principle of gas absorption, *Ind. Eng. Chem.,* 16, 1215, 1924.

63. **Tsunogai, S.,** An estimate of the rate of decomposition of organic matter in the deep water of the Pacific Ocean, in *Biological Oceanography of the Northern North Pacific Ocean,* Takenouti, A. Y., Ed., Idemitsu Shoten, Tokyo, 1972, 517.

64. **Seki, H., Takahashi, M., Hara, Y., and Ichimura, S.,** Dynamics of dissolved oxygen during algal bloom in Lake Kasumigaura, Japan, *Water Res.,* 14, 179, 1980.

65. **Kuroiwa, K., Ogawa, Y., Seki, H., and Ichimura, S.,** Dynamics of dissolved oxygen in a hypereutrophic lake: Lake Kasumigaura, Japan, *Water Air Soil Pollut.,* 12, 255, 1979.

66. **Seki, H., Tsuji, T., and Hattori, A.,** Effect of zooplankton grazing on the formation of the anoxic layer in Tokyo Bay, *Est. Coast. Mar. Sci.,* 2, 145, 1974.

173

67. Pamatmat, M. M., Oxygen consumption by the seabed IV. Shipboard and laboratory experiments, *Limnol. Oceanogr.*, 16, 536, 1971.

68. Mortimer, C. H., The exchange of dissolved substances between mud and water in lakes, *J. Ecol.*, 29, 280, 1941.

69. ZoBell, C. E., The effect of oxygen tension on the rate of oxidation of organic matter in sea water by bacteria, *J. Mar. Res.*, 3, 211, 1940.

70. Seki, H., Microbial respiration in marine environments during summer, *La Mer*, 11, 147, 1973.

71. ZoBell, C. E., Microbial and environmental transitions in estuaries, in *Estuarine Microbial Ecology*, Stevenson, L. H. and Co well, R. R., Eds., University of South Carolina Press, Charleston, 1973, 9.

72. Naumann, E., Über das Neuston des Süsswassers, *Biol. Zentralblat*, 37, 98, 1917.

73. Zaitsev, Yu. P., *Marine Neustonology*, Isr. Program Sci. Trans., Jerusalem, 1971, 207.

74. Hempel, G. and Weikert, H., The neuston of the subtropical and boreal North-eastern Atlantic Ocean. A review, *Mar. Biol.*, 13, 70, 1972.

75. Tsyban, A. V., Marine bacterioneuston, *J. Oceanogr. Soc. Jpn.*, 27, 56, 1971.

76. Tsyban, A. V. and Teplinskaya, N. G., Microbial population of the northwestern Pacific waters, in *Biological Oceanography of the Northern North Pacific Ocean*, Takenouti, A. Y., Ed., Idemitsu Shoten, Tokyo, 1972, 541.

77. Tsiban, A. V., Bacterioplankton and bacterioneuston in the north-eastern part of the Pacific Ocean, *Tezisy Dokl. II Sjezda VGBO, Kishinev*, 92, 1970.

78. Nishizawa, S. and Riley, G. A., Research in particulate materials suspended in sea water, in Proc. 1st National Coast. Shallow Water Res. Conf., Gorsline, D. S., Ed., 1962, 897.

79. Riley, G. A., Particulate and organic matter in sea water, *Adv. Mar. Biol.*, 8, 1, 1970.

80. Pérès, J. M., *Océanographie Biologique et Biologie Marine, Vol. I. La vie Bentique*, Presses Universitaires de France, Paris 1961.

81. Yoshida, T., Studies on the distribution and drift of the floating seaweeds, *Bull. Tohoku Reg. Fish. Res. Lab.*, 23, 141, 1963.

82. Zenkevich, L. A., *Biology of USSR Seas, Their Fauna and Flora*, Uchpedgiz, Moscow, 1956.

83. Strickland, J. D. H., Production of organic matter in the primary stages of the marine food chain, in *Chemical Oceanography, Volume I*, Riley, J. P. and Skirrow, G., Eds., Academic Press, New York, 1965, 477.

84. Stearn, A. E. and Eyring, H., Pressure and rate processes, *Chem. Rev.*, 29, 509, 1941.

85. Brown, D. E., Johnson, F. H., and Marsland, D. A., The pressure-temperature relations of bacterial luminescence, *J. Cell. Comp. Physiol.*, 20, 151, 1942.

86. ZoBell, C. E., Ecology of sulfate reducing bacteria, *Procedures Monthly*, 22, 12, 1958.

87. ZoBell, C. E. and Johnson, F. H., The influence of hydrostatic pressure on the growth and viability of terrestrial and marine bacteria, *J. Bacteriol.*, 57, 179, 1949.

88. ZoBell, C. E. and Morita, R. Y., Deep-sea bacteria, *J. Bacteriol.*, 73, 563, 1957.

89. ZoBell, C. E., Pressure effects on morphology and life processes of bacteria, in *High Pressure Effects on Cellular Processes*, Zimmerman, A. M., Ed., Academic Press, New York, 1970, 85.

90. Jannasch, H. W., Eimhje len, K., Wirsen, C. O., and Farmanfarmaian, A., Microbial degradation of organic matter in the deep sea, *Science*, 171, 672, 1971.

91. Jannasch, H. W. and Wirsen, C. O., Deep-sea micro-organisms: in situ response to nutrient enrichment, *Science*, 180, 641, 1973.

92. Jannasch, H. W. and Wirsen, C. O., Retrieval of concentrated and undecompressed microbial populations from the deep sea, *Appl. Environ. Microbiol.*, 33, 642, 1977.

93. Schwarz, J. R. and Colwell, R. R., Heterotrophic activity of deep-sea sediment bacteria, *Appl. Microbiol.*, 30, 639, 1975.

94. Jannasch, H. W., Wirsen, C. O., and Taylor, C. D., Studies on undecompressed microbial populations from the deep sea, *Appl. Environ. Microbiol.*, 32, 360, 1976.

95. Seki, H., Wada, E., Koike, I., and Hattori, A., Evidence of high organotrophic potentiality of bacteria in the deep ocean, *Mar. Biol.*, 26, 1, 1974.

96. Sieburth, J. M. and Dietz, A. S., Biodeterioration in the sea and its inhibition, in *Effect of the Ocean Environment on Microbial Activities*, Colwell, R. R. and Morita, R. Y., Eds., University Park Press, Baltimore, 1974, 318.

97. Wada, E., Koike, I., and Hattori, A., Nitrate metabolism in abyssal waters, *Mar. Biol.*, 29, 119, 1975.

98. Schwarz, J. R., Yayanos, A. A., and Colwell, R. R., Metabolic activities of the intestinal microflora of a deep-sea invertebrate, *Appl. Environ. Microbiol.*, 31, 46, 1976.

99. Morita, R. Y., Deep-sea microbial energetics, *Sarsia*, 64, 9, 1979.

100. Novitsky, J. A. and Morita, R. Y., Morphological characterization of small cells resulting from nutrient starvation of a psychrophilic marine vibrio, *Appl. Environ. Microbiol.*, 32, 617, 1976.

101. Vinogradov, M. E., Feeding of the deep-sea zooplankton, *Rapp. Proc. Verb. Cons. Perm. Int. Explor. Mer*, 153, 114, 1962.

102. Knight-Jones, E. W. and Morgan, E., Responses of some marine animals to changes in hydrostatic pressure, *Oceanogr. Mar. Biol. Ann. Rev.,* 4, 267, 1966.

103. Winberg, G. G., Rate of metabolism and food requirements of fishes, (in Russian), *Nauchn. Tr. Belorusskovo Gosudarstvennovo Univ. Imeni V. I. Lenina, Minsk,* 1956, 253.

104. Parsons, T. R., LeBrasseur, R. J., and Fulton, J. D., Some observations on the dependence of zooplankton grazing on the cell size and concentration of phytoplankton blooms, *J. Oceanogr. Soc. Jpn.,* 23, 10, 1967.

105. Honjo, S., Sedimentation of materials in the Sargasso Sea at a 5,367 m deep station, *J. Mar. Res.,* 36, 469, 1978.

106. McCave, I. N., Vertical flux of particles in the ocean, *Deep Sea Res.,* 22, 491, 1975.

107. Smayda, T. J., Some measurements of the sinking rate of fecal pellets, *Limnol. Oceanogr.,* 14, 621, 1969.

108. Fowler, S. W. and Small, L. F., Sinking rates of euphausiid fecal pellets, *Limnol. Oceanogr.,* 17, 293, 1972.

109. Wiebe, P. H., Boyd, S. H., and Winget, C., Particulate matter sinking to the deep-sea floor at 2000 m in the Tongue of the Ocean, Bahamas, with a description of a new sedimentation trap, *J. Mar. Res.,* 34, 341, 1976.

110. Honjo, S. and Roman, M. R., Marine copepod fecal pellets: production, preservation and sedimentation, *J. Mar. Res.,* 36, 45, 1978.

111. Knauer, G. A., Martin, J. H., and Bruland, K. W., Fluxes of particulate carbon, nitrogen, and phosphorus in the upper water column of the northeast Pacific, *Deep Sea Res.,* 26, 97, 1979.

112. Tanoue, E. and Handa, N., Vertical transportation of organic materials in the northern North Pacific by sediment trap experiment I. Fatty acid composition, *J. Oceanogr. Soc. Jpn.,* 36, 231, 1981.

113. Tanoue, E. and Handa, N., Vertical transportation of organic materials in the northern North Pacific by sediment trap experiment II. Monosaccharide composition, *J. Oceanogr. Soc. Jpn.,* in press.

114. Tanoue, E. and Handa, N., Vertical transportation of organic materials in the northern North Pacific by sediment trap experiment III. Vertical flux of organic matter and its elementary composition, *J. Oceanogr. Soc. Jpn.,* in press.

115. Honjo, S., Material fluxes and modes of sedimentation in the mesopelagic and bathypelagic zones, *J. Mar. Res.,* 38, 53, 1980.

116. Riley, G. A., Oxygen, phosphate and nitrate in the Atlantic Ocean, *Bull. Bingham Oceanogr. Coll.,* 13, 1, 1951.

117. Sokolova, M. N., Trophic structure of deep-sea macrobenthos, *Mar. Biol.,* 16, 1, 1972.

118. Belyayev, G. M. and Vinogradova, N. G., Quantitative distribution of the bottom fauna in the northern half of the Indian Ocean, *Oceanology,* 136, 35, 1961.

119. Jackson, T. A., Humic matter in natural waters and sediments, *Soil Sci.,* 119, 56, 1975.

120. Degens, E. T., Reuter, J. H., and Shaw, N. F., Biochemical compounds in offshore California sediments and sea waters, *Geochim. Cosmochim. Acta,* 28, 45, 1964.

121. Jannasch, H. W., Growth of marine bacteria at limiting concentrations of organic carbon in seawater, *Limnol. Oceanogr.,* 12, 264, 1967.

122. Hylemon, P. B., Wells, J. S., Jr., Krieg, N. R., and Jannasch, H. W., The genus *Spirillum*: a taxonomic study, *Int. J. Syst. Bacteriol.,* 23, 340, 1973.

123. Stove, J. L. and Stanier, R. Y., Cellular differentiation in stalked bacteria, *Nature (London),* 196, 1189, 1962.

124. Jannasch, H. W., *Caulobacter* sp. in sea water, *Limnol. Oceanogr.,* 5, 432, 1960.

125. London, J., *Thiobacillus intermedius* nov.sp.: a novel type of facultative autotroph, *Arch. Mikrobiol.,* 46, 329, 1963.

126. Starkey, R. L., Cultivation of organisms concerned in the oxidation of thiosulphate, *J. Bacteriol.,* 28, 365, 1934.

127. Taylor, B. F. and Hoare, D. S., New facultative *Thiobacillus* and a reevaluation of the heterotrophic potential of *Thiobacillus novellus, J. Bacteriol.,* 100, 487, 1969.

128. Tuttle, J. H. and Jannasch, H. W., Occurrence and types of thiobacillus-like bacteria in the sea, *Limnol. Oceanogr.,* 17, 532, 1972.

129. Tuttle, J. H. and Jannasch, H. W., Microbial utilization of thiosulfate in the deep sea, *Limnol. Oceanogr.,* 21, 697, 1976.

130. London, J. and Rittenberg, S. C., *Thiobacillus perometabolis* nov. sp., a nonautotrophic thiobacillus, *Arch. Mikrobiol.,* 59, 218, 1967.

131. Tilton, R. C., Cobet, A. B., and Jones, G. E., Marine thiobacilli I. Isolation and distribution, *Can. J. Microbiol.,* 13, 1521, 1967.

132. ZoBell, C. E., Substratum. Bacteria, fungi and blue-green algae, in *Marine Ecology, Volume I,* part 3, Kinne, O., Ed., Wiley-Interscience, London, 1972, 1251.

133. Bader, R. G., Hood, D. W., and Smith, J. B., Recovery of dissolved organic matter in seawater and organic sorption by particulate material, *Geochim. Cosmochim. Acta,* 19, 236, 1960.

134. Chave, K. E., Carbohydrates: association with organic matter in surface seawater, *Science*, 148, 1723, 1965.

135. Degens, E. T. and Matheja, J., Molecular mechanisms on interactions between oxygen co-ordinated metal polyhedra and biochemical compounds, *Woods Hole Oceanogr. Inst. Tech. Rep. Ref.*, 67, , 1967.

136. Sutcliffe, W. H., Baylor, E. R., and Menzel, D. W., Sea surface chemistry and Langmuir circulation, *Deep-Sea Res.*, 10, 233, 1963.

137. Carlucci, A. F. and Williams, P. M., Concentration of bacteria from seawater by bubble scavenging, *J. Cons. Int. Explor. Mer* 30, 28, 1965.

138. Barber, R. T., Dissolved organic carbon from deep water resists microbial oxidation, *Nature (London)*, 220, 274, 1968.

139. Antia, N. J., Effects of temperature on the darkness survival of marine microplanktonic algae, *Mar. Ecol.*, 3, 41, 1976.

140. Antia, N. J. and Cheng, J. Y., The survival of axenic cultures of marine planktonic algae from prolonged exposure to darkness at 20°C, *Phycologia*, 9, 179, 1970.

141. Fenchel, T., Studies on the decomposition of organic detritus derived from turtle grass *Thalassia testudinum*, *Limnol. Oceanogr.*, 15, 14, 1970.

142. Matsuo, S., Yamamoto, H., Nakano, H., and Seki, H., Impact of nutrient enrichment in a water-chestnut ecosystem at Takahama-iri Bay of Lake Kasumigaura, Japan III. Degradation of water-chestnut, *Water Air Soil Pollut.*, 12, 511, 1979.

143. Yamamoto, H. and Seki, H., Impact of nutrient enrichment in a waterchestnut ecosystem at Takahama-iri Bay of Lake Kasumigaura, Japan IV. Population dynamics of secondary producers as indicated by chitin, *Water Air Soil Pollut.*, 12, 519, 1979.

144. Society of American Bacteriologists, *Manual of Microbiological Methods*, McGraw-Hill, New York, 1957, 315.

145. ZoBell, C. E. and Rittenterg, S. C., The occurrence and characteristic of chitinoclastic bacteria in the sea, *J. Bacteriol.*, 35, 275, 1937.

146. Hood, M. A. and Meyers, S. P., The biology of aquatic chitinoclastic bacteria and their chitinoclastic activities, *La Mer*, 11, 213, 1973.

147. Seki, H., Microbiological studies on the decomposition of chitin in marine environment IX. Rough estimation of chitin decomposition on in the ocean, *J. Oceanogr. Soc. Jpn.*, 21, 253, 1965.

148. Seki, H., Microbiological studies on the decomposition of chitin in marine environment X. Decomposition of chitin in marine sediments, *J. Oceanogr. Soc. Jpn.*, 21, 261, 1965.

149. Chan, J. C., The Occurrence, Taxonomy and Activity of Chitinoclastic Bacteria from Sediment Water, and Fauna of Puge Sound, Ph.D. thesis, University of Washington, Seattle, 1970, 312.

150. Seki, H., Ecological studies on the lipolytic activity of microorganisms in the sea of Aburatsubo Inlet, *Rec. Oceanogr. Works Jpn.*, 9, 75, 1967.

151. Williams, P. M., Organic acids in Pacific Ocean waters, *Nature (London)*, 189, 219, 1961.

152. Brockerhoff, H., Yurkowski, M., Hoyle, R. J., and Ackman, R. G., Fatty acid distribution in lipids of marine plankton, *J. Fish. Res. Bd. Can.*, 21, 1379, 1964.

153. Kayama, M., Fatty acid metabolism of fishes, *Bull. Jpn. Soc. Sci. Fish.*, 30, 647, 1964.

154. Peppler, H. J., Bacterial utilization of pure fats and their components, *J. Bacteriol.*, 42, 288, 1941.

155. ZoBell, C. E., *Marine Microbiology*, Chronica Botanica, Waltham, Mass., 1946, 240.

156. Kusnezow, S. I., *Die Rolle der Mikroorganismen im Stoffkreislauf der Seen*, VEB Deutscher Verlag der Wissenschaften, Berlin 1959, 301.

157. Harold, F. M., Conservation and transformation of energy by bacterial membranes, *Bacteriol. Rev.*, 36, 172, 1972.

158. Hellebust, J. A., The uptake and utilization of organic substances by marine phytoplankters, in *Organic Matter in Natural Waters*, Hood, D. W., Ed., University of Alaska, Fairbanks, 1970, 225.

159. Wright, R. T. and Hobbie, J. E., Use of glucose and acetate by bacteria and algae in aquatic ecosystems, *Ecology*, 47, 447, 1966.

160. Seki, H., Nakai, T., and Otobe, H., Regional differences on turnover rate of dissolved materials in the Pacific Ocean at summer of 1971, *Arch. Hydrobiol.*, 71, 79, 1972.

161. Seki, H., Nakai, T., and Otobe, H., Turnover rate of dissolved materials in the Philippine Sea at winter of 1973, *Arch. Hydrobiol.*, 73, 238, 1974.

162. Seki, H., Yamaguchi, Y., and Ichimura, S., Turnover rate of dissolved organic materials in a coastal region of Japan at summer stagnation period of 1974, *Arch. Hydrobiol.*, 75, 297, 1975.

163. Handa, N., Organic materials, in *Marine Biochemistry*, Hattori, A., Ed., University of Tokyo Press, 1973, 65.

164. Seki, H., Terada, T., and Ichimura, S., Steady-state oscillation of uptake kinetics by microorganisms in mesotrophic and eutrophic watermasses, *Arch. Hydrobiol.*, 88, 219, 1980.

165. Seki, H., MacIsaac, E. A., and Stockner, J. G., The turnover rate of dissolved organic material in waters used by anadromous Pacific salmon on their return to Great Central Lake on Vancouver Island, British Columbia, Canada, *Arch. Hydrobiol.*, 88, 58, 1980.

166. Barraclough, W. E. and Robinson, D., The fertilization of Great Central Lake III. Effect on juvenile sockeye salmon, *Fish. Bull. U.S.*, 70, 37, 1971.

167. Barraclough, W. E., LeBrasseur, R. J., and McAllister, C. D., Lake fertilization: an experimental approach to manipulating food chains, in Spec. Symp. Mar. Sci., Hong Kong, 1973, 11.

168. LeBrasseur, R. J. and Kennedy, O. D., The fertilization of Great Central Lake II. Zooplankton standing stock, *Fish. Bull. U.S.*, 70, 25, 1972.

169. LeBrasseur, R. J., McAllister, C. D., Barraclough, W. E., Kennedy, O. D., Manzer, J., Robinson, D., and Stephens, K., Enhancement of sockeye salmon *(Oncorhynchus nerka)* by lake fertilization in Great Central Lake: summary report, *J. Fish. Res. Bd. Can.*, 35, 1580, 1978.

170. Parsons, T. R., McAllister, C. D., LeBrasseur, R. J., and Barraclough, W. E., The use of nutrients in the enrichment of sockeye salmon nursery lakes (A preliminary report), in *Marine Pollution and Sea Life*, Ruivo, M., Ed., Fishing News Ltd., London, 1972, 1.

171. Parsons, T. R., Stephens, K., and Takahashi, M., The fertilization of Great Central Lake I. Effect on primary production, *Fish. Bull. U.S.*, 70, 13, 1972.

172. Stockner, J. G., Stephens, K., and Shortreed, K. S., Artificially induced upwelling in Great Central Lake, British Columbia, *Fish. Mar. Serv. Tech. Rep.*, 803, 1, 1975.

173. Stockner, J. G., Lake fertilization as a means of enhancing sockeye salmon populations: The state of the art in the Pacific Northwest, *Fish. Mar. Serv. Tech. Rep.*, 740, 1, 1977.

174. Seki, H., unpublished data, 1979.

175. Seki, H., Shortreed, K. S., and Stockner, J. G., Turnover rate of dissolved organic materials in glacially-oligotrophic and dystrophic lakes in British Columbia, Canada, *Arch. Hydrobiol.*, 90, 210, 1980.

176. Ohle, W., Sulfat als "katalysator" des limnischen Stoffkreislaufes, *Vom Wasser.*, 21, 13, 1954.

177. Seki, H., Enrichment of the Pacific waters and steady-state oscillation of uptake kinetics by microorganisms, in Proc. 15th Pac. Sci. Congr., in press.

178. Burnison, B. K. and Morita, R. Y., Heterotrophic potential for amino acid uptake in a naturally eutrophic lake, *Appl. Microbiol.*, 27, 488, 1974.

179. Gocke, K., Heterotrophic activity, in *Microbial Ecology of a Brackish Water Environment*, Rheinheimer, G., Ed., Springer-Verlag, Berlin, 1977, 198.

180. Hamilton, R. D. and Preslan, J. E., Observations on heterotrophic activity in the eastern tropical Pacific, *Limnol. Oceanogr.*, 15, 395, 1970.

181. Hoppe, H. G., Relations between active bacteria and heterotrophic potential in the sea, *Neth. J. Sea Res.*, 12, 78, 1978.

182. Parsons, T. R., Thomas, W. H., Seibert, D., Beers, J. R., Gillespie, P., and Bawden, C., The effect of nutrient enrichment on the plankton community in enclosed water columns, *Int. Revue Ges. Hydrobiol.*, 62, 565, 1977.

183. Seki, H., unpublished data, 1980.

184. Seki, H., Whitney, F., and Wong, C. S., Uptake kinetics of dissolved organic materials in a marine ecosystem with experimental precedence of the detritus food chain, *Arch. Hydrobiol.*, in press.

185. Seki, H., Aoshima, N., and Wong, C. S., Microbial readjustment to new balance after influx change of organic material in marine dysphotic layer, *Arch. Hydrobiol.*, in press.

186. Seki, H. and Nakano, H., Production of bacterioplankton with special reference to dynamics of dissolved organic matter in a hypereutrophic lake, *Kiel. Meeresforsch.*, in press.

187. Seki, H., Takahashi, M., and Ichimura, S., Impact of nutrient enrichment in a waterchestnut ecosystem at Takahama-iri Bay of Lake Kasumigaura, Japan. I. Nutrient influx and phytoplankton bloom, *Water Air Soil Pollut.*, 12, 383, 1979.

188. Seki, H., Aspetti biologici della attuale eutrofizzazione nella Baia de Tokyo Giappone, Seminario Internationale sui Fenomeni di Eutrofizzazione lungo le coste dell'Emilia-Romagna, 1977, 108.

189. Oxley, T. A., Allsopp, D., and Becker, G., Eds., *Biodeterioration*, Imprint (Print & Design) Ltd., Birmingham, 1980, 375.

190. Redfield, A. C., Ketchum, B. H., and Richards, F. A., The influence of organisms on the composition of sea water, in *The Sea. Ideas and Observations on Progress in the Study of the Seas*, Vol. 2, Hill, M. N., Ed., Interscience, New York, 1963, 26.

191. Stanier, R. Y., Doudoroff, M., and Adelberg, E. A., *The Microbial World*, Prentice-Hall, Englewood Cliffs, N.J., 1963, 753.

192. Sorokin, Yu. I., Decomposition of organic matter and nutrient regeneration, in *Marine Ecology, Volume IV*, Kinne, O., Ed., John Wiley & Sons, New York, 1978, 501.

193. Jannasch, H. W. and Wirsen, C. O., Chemosynthetic primary production at east Pacific sea floor spreading centers, *BioScience*, 29, 592, 1979.

194. **Karl, D. M., Wirsen, C. O., and Jannasch, H. W.,** Deep-sea primary production at the Galápagos hydrothermal vents, *Science,* 207, 1345, 1980.
195. **Wong, C. S.,** Carbon dioxide — a global environmental problem into the future, *Mar. Pollut. Bul..,* 9, 264, 1978.
196. **Koike, I. and Hattori, A.,** Denitrification and ammonia formation in anaerobic coastal sediments, *Appl. Environ. Microbiol.,* 35, 278, 1978.
197. **Koike, I. and Hattori, A.,** Simultaneous determinations of nitrification and nitrate reduction in coastal sediments by a ^{15}N dilution technique, *Appl. Environ. Microbiol.,* 35, 853, 1978.
198. **Vollenweider, R. A.,** Advances in defining critical loading levels for phosphorus in lake eutrophication, *Mem. Ist. Ital. Idrobiol.,* 33, 53, 1976.
199. **Duursma, E. K.,** Dissolved organic carbon, nitrogen and phosphorus in the sea, *Neth. J. Sea Res.,* 1, 1, 1961.
200. **Olson, J. S.,** Energy storage and the balance of producers and decomposers in ecological systems, *Ecology,* 44, 322, 1963.
201. **Skopintsev, B. A.,** Some aspects of the distribution and composition of organic matter in the waters of the ocean, *Oceanology* 6, 41, 1966.
202. **Skopintsev, B. A., Timofeeva S. N., and Vershinina, O. A.,** Organic carbon in the equatorial and southern Atlantic and in the Mediterranean, *Oceanology,* 6, 251, 1966.
203. **Williams, P. M., Oeschger, H., and Kinney, P.,** Natural radiocarbon activity of the dissolved organic carbon in the North-east Pacific Ocean, *Nature (London),* 224, 256, 1969.

INDEX

179

A

Abiotic organic materials, in surface films, 49
Abiotic oxidation, rate of, 46
Abiotic particles, sinking of, 5
Absorption spectra, photosynthetic organisms, 7, 19
Aburatsubo Inlet, 29, 46—47, 74, 115, 145
Acartia clausi, 45
Accumulation, see also Concentration
  cellulose, 101
  chitin, 101
  local, metabolizable organic materials, 54—55
  organic solutes, at liquid-gas interface, 61—62
  substrate, 104
Acetic acid, 97, 101, 161—168
Acetylamino units, 100
Acetyl-CoA, 105
N-Acetylglucosamine, 100—101
N-Acetylglucosaminodase, 101
Achromobacter aquamarinus, 58, 92
Achromobacter fischeri, 52, 89
Achromobacter harveyi, 52, 89
Achromobacter sp., 37, 58, 92, 101, 103
Achromobacter thalassius, 52, 89
Acid lakes, 40
Acids, see also specific types by name
  inflowing, pH and, 40
Actinomycetes, 97, 101
Active transport, see Transport, active
Adenosine triphosphate, 5, 13, 15, 40, 44, 103, 105, 113—115, 117, 141
Adriatic Sea, 107
Adsorption, microorganisms and organic materials, on solid surfaces, 60—61
Advection, 118
Advection — diffusion model, oxygen diffusion and organic matter decomposition rate, 43—44, 88
Aeration, 43, 61, 98
  rate of, 43
Aerobes, 42, 44, 46—47, 58, 91, 105
  Brock's classification, 42
  electron transport in, 47
  facultative, 42
  heterotrophic, 58
  obligate, 42, 44
Aerobic bacteria, 59, 95—97, 116
  obligate, 116
  respiratory rates, 95
Aerobic benthos, 46
Aerobic energy-generating processes, minimum oxygen concentration for, 46
Aerobic environment, conditions in, 3, 42, 44—46, 57, 115
  micro-, 46
  microorganisms actively modifying, 42, 44—45
Aerobic microbial flora, 46
Aerobic processes, 118

Aerobic treatment, sewage and waste waters, 120
Aeromonas sp., 101
Agarum sp., 7
Age, Earth, 1
Aggregates, 3, 12—13, 28, 39, 45—46, 49—50, 54, 57, 61—62, 87, 96, 104
Agricultural wastes, eutrophication caused by, 13
Alanine, 107, 134, 153—168
Alberni Inlet, 109, 132, 166
Alcohols, 105
Aleutian Gut, 74
Algae, 2, 4—8, 19—20, 40, 44—45, 50, 68, 98, 113, 116, 119—120, 122, 126, 140, 149—150
  blue-green, 2, 4, 40, 45, 50, 98, 113, 116, 119—120, 126, 140, 149—150
  brown, 7—8
  distribution of, 7, 20
  eucaryotic, 40
  evolution and, 2, 4
  floating, 50
  green, 4, 8, 122
  hyponeustonic, 50
  light reactions in, 5
  primary production and, 5—8, 19
  red, 4, 8
Algal bloom, 44—45, 98, 113, 119—120, 126, 140, 149—151
Alkaline waters, 39—40
Allochthonous humus, annual input, 121
Allochthonous organic matter, 39
Allochthonous organisms, 50, 52—53
Aluminum, 61
Amino acid, 1, 11, 36, 41, 56—57, 95, 105, 107—110, 116—118, 120, 134, 169
  deamination, 41
  turnover rate and time, 110, 120, 134, 169
Amino acid carboxylase, 40
Amino acid deaminase, 40
Ammonia, 16, 41, 116
Ammonification, 116
Ammonifiers, 91
Ammonium, 1—2, 14, 41, 45, 47—48, 116
Amoebae, 12, 48
Amphipods, 55, 91
Ampullae of Lorenzini, 54
Anabaena spiroides, 44, 116
Anabaena sp., 98
Anaerobes, 1, 3, 17, 42, 46—47, 91, 103
  Brock's classification, 42
  facultative, 42
  obligate, 42
Anaerobic bacteria, 97, 116
Anaerobic electron acceptors, 1—2
Anaerobic environments, conditions in, 3, 42, 44—47, 57, 97, 115—116, 118
Anaerobic microzones, see Microzones, anaerobic
Anaerobic processes, 47, 118
Anaerobic respiration, 47

G

M